Fashion supply chain management using radio frequency identification (RFID) technologies

The Textile Institute and Woodhead Publishing

The Textile Institute is a unique organisation in textiles, clothing and footwear. Incorporated in England by a Royal Charter granted in 1925, the Institute has individual and corporate members in over 90 countries. The aim of the Institute is to facilitate learning, recognise achievement, reward excellence and disseminate information within the global textiles, clothing and footwear industries.

Historically, The Textile Institute has published books of interest to its members and the textile industry. To maintain this policy, the Institute has entered into partnership with Woodhead Publishing Limited to ensure that Institute members and the textile industry continue to have access to high calibre titles on textile science and technology.

Most Woodhead titles on textiles are now published in collaboration with The Textile Institute. Through this arrangement, the Institute provides an Editorial Board which advises Woodhead on appropriate titles for future publication and suggests possible editors and authors for these books. Each book published under this arrangement carries the Institute's logo.

Woodhead books published in collaboration with The Textile Institute are offered to Textile Institute members at a substantial discount. These books, together with those published by The Textile Institute that are still in print, are offered on the Elsevier website at: http://store.elsevier.com/. Textile Institute books still in print are also available directly from the Institute's website at: www.textileinstitutebooks.com

A list of Woodhead books on textiles science and technology, most of which have been published in collaboration with the Textile Institute, can be found towards the end of the contents pages.

Woodhead Publishing Series in Textiles: Number 152

Fashion supply chain management using radio frequency identification (RFID) technologies

Edited by

W. K. Wong and Z. X. Guo

The Textile Institute

ELSEVIER

AMSTERDAM • BOSTON • CAMBRIDGE • HEIDELBERG • LONDON
NEW YORK • OXFORD • PARIS • SAN DIEGO
SAN FRANCISCO • SINGAPORE • SYDNEY • TOKYO
Woodhead Publishing is an imprint of Elsevier

WP
WOODHEAD PUBLISHING

Published by Woodhead Publishing Limited in association with The Textile Institute
Woodhead Publishing is an imprint of Elsevier
80 High Street, Sawston, Cambridge, CB22 3HJ, UK
225 Wyman Street, Waltham, MA 02451, USA
Langford Lane, Kidlington, OX5 1GB, UK

British Library Cataloguing-in-Publication Data
A catalogue record for this book is available from the British Library

Library of Congress Control Number: 2013954937

ISBN 978-0-85709-805-4 (print)
ISBN 978-0-85709-811-5 (online)

For information on all Woodhead Publishing publications visit our website at http://store.elsevier.com/

Typeset by RefineCatch Limited, Bungay, Suffolk

Printed and bound in the United Kingdom

Contents

Contributor contact details

(* = main contact)

Editors

W. K. Wong*
Institute of Textiles and Clothing
The Hong Kong Polytechnic University
Hung Hom
Kowloon, Hong Kong, China

E-mail: calvin.wong@polyu.edu.hk

Z. X. Guo
School of Business
Sichuan University
Chengdu 610065, China

E-mail: zx.guo@alumni.polyu.edu.hk

Chapter 1

W. K. Wong
Institute of Textiles and Clothing
The Hong Kong Polytechnic University
Hung Hom
Kowloon, Hong Kong, China

E-mail: calvin.wong@polyu.edu.hk

Chapter 2

E. Bottani*, A. Rizzi, R. Montanari, M. Bertolini and A. Volpi
Industrial Engineering Department
University of Parma
Viale G.P. Usberti 181/A
43124 Parma, Italy

E-mail: eleonora.bottani@unipr.it; antonio.rizzi@unipr.it; roberto.montanari@unipr.it; massimo.bertolini@unipr.it; andrea.volpi@unipr.it

Chapter 3

S. G. Azevedo*
UNIDEMI – Department of Management and Economics
University of Beira Interior
Pólo IV – Edifício Ernesto Cruz
6200-209 Covilhã, Portugal

E-mail: sazevedo@ubi.pt

P. Prata and P. Fazendeiro
Instituto de Telecomunicações (IT)
Department of Informatics, University of Beira Interior
6200-001 Covilhã, Portugal

E-mail: pprata@di.ubi.pt; fazendeiro@ubi.pt

Chapter 4

E. W. T. Ngai*, D. C. K. Chau, A. Y. M. Chan, B. C. M. Chan, and W. W. S. Wu
Department of Management and Marketing
The Hong Kong Polytechnic University
Hung Hom
Kowloon, Hong Kong, China

E-mail: eric.ngai@polyu.edu.hk

J. K. L. Poon
Hong Kong Community College
The Hong Kong Polytechnic University
Hung Hom
Kowloon, Hong Kong, China

Chapter 5

Z. X. Guo
School of Business
Sichuan University
Chengdu 610065, China

E-mail: zx.guo@alumni.polyu.edu.hk

W. K. Wong*, S. Y. S Leung, J. T. Fan and S. F. Chan
Institute of Textiles and Clothing
The Hong Kong Polytechnic University
Hung Hom
Kowloon, Hong Kong, China

E-mail: calvin.wong@polyu.edu.hk

Chapter 6

M. Teucke* and B. Scholz-Reiter
BIBA – Bremer Institut für Produktion und Logistik GmbH
University of Bremen
Hochschulring 20
D-28359 Bremen, Germany

E-mail: tck@biba.uni-bremen.de; bsr@biba.uni-bremen.de

Chapter 7

W. K. Wong*, S. Y. S. Leung, Z. H. Zeng and P. Y. Mok
Institute of Textiles and Clothing
The Hong Kong Polytechnic University
Hung Hom
Kowloon, Hong Kong, China

E-mail: calvin.wong@polyu.edu.hk

Z. X. Guo
School of Business
Sichuan University
Chengdu 610065, China

E-mail: zx.guo@alumni.polyu.edu.hk

Chapter 8

A. Lui and C. K. Y. Lo*
The Hong Kong Polytechnic University
Hung Hom
Kowloon, Hong Kong, China

E-mail: kwan.yu.lo@polyu.edu.hk

Chapter 9

A. De Marco*, A. C. Cagliano, M. L. Nervo and C. Rafele
Dept. of Management and Production Engineering
Politecnico di Torino
Corso Duca degli Abruzzi 24
Torino 10129, Italy

E-mail: alberto.demarco@polito.it; anna.cagliano@polito.it; mauro.nervo@gmail.com; carlo.rafele@polito.it

Woodhead Publishing Series in Textiles

1 **Watson's textile design and colour. Seventh edition**
Edited by Z. Grosicki
2 **Watson's advanced textile design**
Edited by Z. Grosicki
3 **Weaving. Second edition**
P. R. Lord and M. H. Mohamed
4 **Handbook of textile fibres. Volume 1: Natural fibres**
J. Gordon Cook
5 **Handbook of textile fibres. Volume 2: Man-made fibres**
J. Gordon Cook
6 **Recycling textile and plastic waste**
Edited by A. R. Horrocks
7 **New fibers. Second edition**
T. Hongu and G. O. Phillips
8 **Atlas of fibre fracture and damage to textiles. Second edition**
J. W. S. Hearle, B. Lomas and W. D. Cooke
9 **Ecotextile '98**
Edited by A. R. Horrocks
10 **Physical testing of textiles**
B. P. Saville
11 **Geometric symmetry in patterns and tilings**
C. E. Horne
12 **Handbook of technical textiles**
Edited by A. R. Horrocks and S. C. Anand
13 **Textiles in automotive engineering**
W. Fung and J. M. Hardcastle
14 **Handbook of textile design**
J. Wilson
15 **High-performance fibres**
Edited by J. W. S. Hearle
16 **Knitting technology. Third edition**
D. J. Spencer
17 **Medical textiles**
Edited by S. C. Anand
18 **Regenerated cellulose fibres**
Edited by C. Woodings

19 **Silk, mohair, cashmere and other luxury fibres**
Edited by R. R. Franck
20 **Smart fibres, fabrics and clothing**
Edited by X. M. Tao
21 **Yarn texturing technology**
J. W. S. Hearle, L. Hollick and D. K. Wilson
22 **Encyclopedia of textile finishing**
H-K. Rouette
23 **Coated and laminated textiles**
W. Fung
24 **Fancy yarns**
R. H. Gong and R. M. Wright
25 **Wool: Science and technology**
Edited by W. S. Simpson and G. Crawshaw
26 **Dictionary of textile finishing**
H-K. Rouette
27 **Environmental impact of textiles**
K. Slater
28 **Handbook of yarn production**
P. R. Lord
29 **Textile processing with enzymes**
Edited by A. Cavaco-Paulo and G. Gübitz
30 **The China and Hong Kong denim industry**
Y. Li, L. Yao and K. W. Yeung
31 **The World Trade Organization and international denim trading**
Y. Li, Y. Shen, L. Yao and E. Newton
32 **Chemical finishing of textiles**
W. D. Schindler and P. J. Hauser
33 **Clothing appearance and fit**
J. Fan, W. Yu and L. Hunter
34 **Handbook of fibre rope technology**
H. A. McKenna, J. W. S. Hearle and N. O'Hear
35 **Structure and mechanics of woven fabrics**
J. Hu
36 **Synthetic fibres: Nylon, polyester, acrylic, polyolefin**
Edited by J. E. McIntyre
37 **Woollen and worsted woven fabric design**
E. G. Gilligan
38 **Analytical electrochemistry in textiles**
P. Westbroek, G. Priniotakis and P. Kiekens
39 **Bast and other plant fibres**
R. R. Franck
40 **Chemical testing of textiles**
Edited by Q. Fan
41 **Design and manufacture of textile composites**
Edited by A. C. Long
42 **Effect of mechanical and physical properties on fabric hand**
Edited by H. M. Behery

43 **New millennium fibers**
T. Hongu, M. Takigami and G. O. Phillips
44 **Textiles for protection**
Edited by R. A. Scott
45 **Textiles in sport**
Edited by R. Shishoo
46 **Wearable electronics and photonics**
Edited by X. M. Tao
47 **Biodegradable and sustainable fibres**
Edited by R. S. Blackburn
48 **Medical textiles and biomaterials for healthcare**
Edited by S. C. Anand, M. Miraftab, S. Rajendran and J. F. Kennedy
49 **Total colour management in textiles**
Edited by J. Xin
50 **Recycling in textiles**
Edited by Y. Wang
51 **Clothing biosensory engineering**
Y. Li and A. S. W. Wong
52 **Biomechanical engineering of textiles and clothing**
Edited by Y. Li and D. X-Q. Dai
53 **Digital printing of textiles**
Edited by H. Ujiie
54 **Intelligent textiles and clothing**
Edited by H. R. Mattila
55 **Innovation and technology of women's intimate apparel**
W. Yu, J. Fan, S. C. Harlock and S. P. Ng
56 **Thermal and moisture transport in fibrous materials**
Edited by N. Pan and P. Gibson
57 **Geosynthetics in civil engineering**
Edited by R. W. Sarsby
58 **Handbook of nonwovens**
Edited by S. Russell
59 **Cotton: Science and technology**
Edited by S. Gordon and Y-L. Hsieh
60 **Ecotextiles**
Edited by M. Miraftab and A. R. Horrocks
61 **Composite forming technologies**
Edited by A. C. Long
62 **Plasma technology for textiles**
Edited by R. Shishoo
63 **Smart textiles for medicine and healthcare**
Edited by L. Van Langenhove
64 **Sizing in clothing**
Edited by S. Ashdown
65 **Shape memory polymers and textiles**
J. Hu
66 **Environmental aspects of textile dyeing**
Edited by R. Christie

67 **Nanofibers and nanotechnology in textiles**
Edited by P. Brown and K. Stevens
68 **Physical properties of textile fibres. Fourth edition**
W. E. Morton and J. W. S. Hearle
69 **Advances in apparel production**
Edited by C. Fairhurst
70 **Advances in fire retardant materials**
Edited by A. R. Horrocks and D. Price
71 **Polyesters and polyamides**
Edited by B. L. Deopura, R. Alagirusamy, M. Joshi and B. S. Gupta
72 **Advances in wool technology**
Edited by N. A. G. Johnson and I. Russell
73 **Military textiles**
Edited by E. Wilusz
74 **3D fibrous assemblies: Properties, applications and modelling of three-dimensional textile structures**
J. Hu
75 **Medical and healthcare textiles**
Edited by S. C. Anand, J. F. Kennedy, M. Miraftab and S. Rajendran
76 **Fabric testing**
Edited by J. Hu
77 **Biologically inspired textiles**
Edited by A. Abbott and M. Ellison
78 **Friction in textile materials**
Edited by B. S. Gupta
79 **Textile advances in the automotive industry**
Edited by R. Shishoo
80 **Structure and mechanics of textile fibre assemblies**
Edited by P. Schwartz
81 **Engineering textiles: Integrating the design and manufacture of textile products**
Edited by Y. E. El-Mogahzy
82 **Polyolefin fibres: Industrial and medical applications**
Edited by S. C. O. Ugbolue
83 **Smart clothes and wearable technology**
Edited by J. McCann and D. Bryson
84 **Identification of textile fibres**
Edited by M. Houck
85 **Advanced textiles for wound care**
Edited by S. Rajendran
86 **Fatigue failure of textile fibres**
Edited by M. Miraftab
87 **Advances in carpet technology**
Edited by K. Goswami
88 **Handbook of textile fibre structure. Volume 1 and Volume 2**
Edited by S. J. Eichhorn, J. W. S. Hearle, M. Jaffe and T. Kikutani
89 **Advances in knitting technology**
Edited by K-F. Au
90 **Smart textile coatings and laminates**
Edited by W. C. Smith

91 **Handbook of tensile properties of textile and technical fibres**
Edited by A. R. Bunsell
92 **Interior textiles: Design and developments**
Edited by T. Rowe
93 **Textiles for cold weather apparel**
Edited by J. T. Williams
94 **Modelling and predicting textile behaviour**
Edited by X. Chen
95 **Textiles, polymers and composites for buildings**
Edited by G. Pohl
96 **Engineering apparel fabrics and garments**
J. Fan and L. Hunter
97 **Surface modification of textiles**
Edited by Q. Wei
98 **Sustainable textiles**
Edited by R. S. Blackburn
99 **Advances in yarn spinning technology**
Edited by C. A. Lawrence
100 **Handbook of medical textiles**
Edited by V. T. Bartels
101 **Technical textile yarns**
Edited by R. Alagirusamy and A. Das
102 **Applications of nonwovens in technical textiles**
Edited by R. A. Chapman
103 **Colour measurement: Principles, advances and industrial applications**
Edited by M. L. Gulrajani
104 **Fibrous and composite materials for civil engineering applications**
Edited by R. Fangueiro
105 **New product development in textiles: Innovation and production**
Edited by L.Horne
106 **Improving comfort in clothing**
Edited by G. Song
107 **Advances in textile biotechnology**
Edited by V. A. Nierstrasz and A. Cavaco-Paulo
108 **Textiles for hygiene and infection control**
Edited by B. McCarthy
109 **Nanofunctional textiles**
Edited by Y. Li
110 **Joining textiles: Principles and applications**
Edited by I. Jones and G. Stylios
111 **Soft computing in textile engineering**
Edited by A. Majumdar
112 **Textile design**
Edited by A. Briggs-Goode and K. Townsend
113 **Biotextiles as medical implants**
Edited by M. W. King, B. S. Gupta and R. Guidoin
114 **Textile thermal bioengineering**
Edited by Y. Li
115 **Woven textile structure**
B. K. Behera and P. K. Hari

116 **Handbook of textile and industrial dyeing. Volume 1: Principles, processes and types of dyes**
Edited by M. Clark
117 **Handbook of textile and industrial dyeing. Volume 2: Applications of dyes**
Edited by M. Clark
118 **Handbook of natural fibres. Volume 1: Types, properties and factors affecting breeding and cultivation**
Edited by R. Kozłowski
119 **Handbook of natural fibres. Volume 2: Processing and applications**
Edited by R. Kozłowski
120 **Functional textiles for improved performance, protection and health**
Edited by N. Pan and G. Sun
121 **Computer technology for textiles and apparel**
Edited by J. Hu
122 **Advances in military textiles and personal equipment**
Edited by E. Sparks
123 **Specialist yarn and fabric structures**
Edited by R. H. Gong
124 **Handbook of sustainable textile production**
M. I. Tobler-Rohr
125 **Woven textiles: Principles, developments and applications**
Edited by K. Gandhi
126 **Textiles and fashion: Materials design and technology**
Edited by R. Sinclair
127 **Industrial cutting of textile materials**
I. Viļumsone-Nemes
128 **Colour design: Theories and applications**
Edited by J. Best
129 **False twist textured yarns**
C. Atkinson
130 **Modelling, simulation and control of the dyeing process**
R. Shamey and X. Zhao
131 **Process control in textile manufacturing**
Edited by A. Majumdar, A. Das, R. Alagirusamy and V. K. Kothari
132 **Understanding and improving the durability of textiles**
Edited by P. A. Annis
133 **Smart textiles for protection**
Edited by R. A. Chapman
134 **Functional nanofibers and applications**
Edited by Q. Wei
135 **The global textile and clothing industry: Technological advances and future challenges**
Edited by R. Shishoo
136 **Simulation in textile technology: Theory and applications**
Edited by D. Veit
137 **Pattern cutting for clothing using CAD: How to use Lectra Modaris pattern cutting software**
M. Stott
138 **Advances in the dyeing and finishing of technical textiles**
M. L. Gulrajani

139 **Multidisciplinary know-how for smart textiles developers**
Edited by T. Kirstein
140 **Handbook of fire resistant textiles**
Edited by F. Selcen Kilinc
141 **Handbook of footwear design and manufacture**
Edited by A. Luximon
142 **Textile-led design for the active ageing population**
Edited by J. McCann and D. Bryson
143 **Optimizing decision making in the apparel supply chain using artificial intelligence (AI): From production to retail**
Edited by W. K. Wong, Z. X. Guo and S. Y. S. Leung
144 **Mechanisms of flat weaving technology**
V. V. Choogin, P. Bandara and E. V. Chepelyuk
145 **Innovative jacquard textile design using digital technologies**
F. Ng and J. Zhou
146 **Advances in shape memory polymers**
J. Hu
147 **Design of clothing manufacturing processes: A systematic approach to planning, scheduling and control**
J. Gersak
148 **Anthropometry, apparel sizing and design**
D. Gupta and N. Zakaria
149 **Silk: Processing, properties and applications**
Edited by K. Murugesh Babu
150 **Advances in filament yarn spinning of textiles and polymers**
Edited by D. Zhang
151 **Designing apparel for consumers: The impact of body shape and size**
Edited by M.-E. Faust and S. Carrier
152 **Fashion supply chain management using radio frequency identification (RFID) technologies**
Edited by W. K. Wong and Z. X. Guo
153 **High performance textiles and their applications**
Edited by C. Lawrence
154 **Protective clothing: Managing thermal stress**
Edited by F. Wang and C. Gao
155 **Composite nonwoven materials**
Edited by D. Das and B. Pourdeyhimi
156 **Functional finishes for textiles**
Edited by R. Paul
157 **Assessing the environmental impact of textiles and the clothing supply chain**
S. S. Muthu
158 **Braiding technology for textiles**
Y. Kyosev
159 **Principles of colour appearance and measurement**
A. K. R. Choudhury

1
The role of radio frequency identification (RFID) technologies in the textiles and fashion supply chain: an overview

W. K. WONG, The Hong Kong Polytechnic University, Hong Kong and Z. X. GUO, Sichuan University, China

DOI: 10.1533/9780857098115.1

Abstract: Barcode and radio frequency identification (RFID) technology have been widely applied in automatic identification and tracking throughout the textiles and fashion supply chain. This chapter will first compare the differences between these technologies and discuss how RFID technology can perform better than barcode technology in various aspects. The fundamentals of RFID technology, the architecture of an RFID system and an overview of the application of RFID technology in the textiles and fashion supply chain will be described.

Key words: barcode, radio frequency identification (RFID), system architecture, textiles and fashion supply chain.

1.1 Introduction

The apparel industry is one of the most important sectors of the economy, creating jobs and products that meet fundamental human needs. The supply chain of the apparel industry is highly complex due to a number of distinct industrial features, which include short product life-cycles, a wide product range and volatile customer demand. It is becoming an increasing trend to employ radio frequency identification (RFID) technology to identify and track individual products in the apparel supply chain.

A supply chain is a set of policies, processes, management actions, and technologies that collectively forecast, acquire and deliver products and services to meet the identified needs of a company and/or customer (Shepard, 2005). Supply chain management (SCM) is the management of a network of interconnected business processes involved in a supply chain for the purpose of creating value for customers and stakeholders. It spans all movement and storage of raw materials, work-in-process inventory and finished goods. Effective SCM is crucial to lower operating costs and improve the competitiveness of businesses. Good SCM depends on the availability of accurate and timely data about various activities in the supply chain. These include progress in meeting production schedules, current inventory levels and the location of material. Data acquisition is thus crucial. For example, early detection of a shipment delay by rail may mean that it can still make its deadline by trucking it, while later detection of this problem may mean that some critical parts may have to be delivered by air, a much more expensive option.

In recent years, automatic identification systems have become popular in many manufacturing and service industries as well as the logistics industry, which aim to improve data acquisition processes and to capture data in a more timely and accurate manner. This in turn gives management and decision-makers more time to recognize potential problems and make efficient decisions. The application of automatic identification systems involves a broad range of supply chain operations, such as:

- item identification and tracking;
- manufacturing;
- retailing;
- transportation;
- warehousing; and
- payment transactions.

Technologies such as barcode, RFID, signature capture and magnetic stripe, the former two being most commonly used in logistics and SCM, are indeed already in use.

1.2 From barcode to RFID technology

A barcode is an optical machine-readable representation of data relating to the object to which it is attached. The invention of the barcode is usually credited to two graduate students at Drexel Institute of Technology, Norman Joseph Woodland and Bernard Silver, who pioneered the concept and filed the first patent on barcode technology on 20 October 1949. The first application of industrial barcode technology was in the labeling of railroad cars in the 1960s, but it was not commercially successful until the now ubiquitous Universal Product Code (UPC) barcode. This was developed and used to automate the grocery checkout process, the first occasion of which has been credited on a packet of chewing gum in Troy, Ohio, in 1974. The successful application of barcode technology revolutionized inventory tracking and management in the retail industry.

Nowadays, barcodes are used almost universally. Airlines rely on barcodes to track passenger luggage to reduce the chance of loss. Warehouses rely on them to track the movements of materials and parts. The express industry relies on them to track mails and parcels. Barcodes are attached to each book to make the borrowing, returning and stocktaking of books easier and faster. Researchers have also looked into insects' mating habits, such as by placing tiny barcodes on individual bees. NASA depends on barcodes to monitor thousands of heat tiles that need to be replaced after every space shuttle trip. Barcodes even appear on humans! Stamping barcodes onto fashion models helps designers to coordinate their fashion shows.

Many different types of barcode technologies have been developed since the emergence of the UPC barcode, which can be classified mainly into two categories:

1. linear barcode;
2. two-dimensional (2D) barcode.

Each category comprises multiple barcode types. The linear barcode is made up of parallel lines and spaces of various widths that create specific patterns, which typically encodes alpha-numerical strings up to maximum of about 20 characters. The 2D barcode is made up of rectangles, dots, hexagons and other geometric patterns in two dimensions, which can represent more data per unit area than the linear barcode.

Although it is well developed and its applications are wide-ranging, barcode technology does have some limitations. For example, we are not able to change the data on a barcode once it is printed. In addition, barcode scanning devices are designed to operate over short distances. It is these limitations which RFID technology is capable of overcoming.

RFID is a generic term used to describe technologies that involve the use of a wireless non-contact system. This utilizes radio waves to transfer data from an RFID tag attached to an object, for the purposes of automatic identification and tracking. Unlike barcode technology, RFID tags can be read from up to tens of meters away and beyond the lines of sight of the reader. Although the history of radio frequency engineering can be traced back to 1864, when James Clerk Maxwell predicted the existence of electromagnetic waves through his famous equations, RFID was not invented until 60 years later. The first RFID application was developed in conjunction with radar technology for the Identification Friend or Foe (IFF) system used during World War II. The IFF system was developed in a secret British project led by Scottish inventor and physicist Sir Robert Alexander Watson-Watt. In this system, active and battery-powered RFID tags were placed on each British plane. When planes received radio frequency signals from radar stations on the ground, they broadcast signals back in the opposite direction which identified the plane as friendly. A passive radio transponder with memory, invented by Mario W. Cardullo and patented on 23 January 1973, is usually viewed as the first patent for an RFID tag. However, the first patent to be associated with the abbreviation RFID would not appear for a further ten years, this being granted to Charles Walton.

The applications of RFID technology have increased rapidly since its emergence. Nowadays, RFID can be used almost anywhere where a unique identification system is needed. The major areas of RFID applications involve:

- access control;
- personnel and product identification;
- fleet management;
- vehicle identification;
- production line monitoring;

- passport security;
- mail and shipping; and
- baggage handling.

1.3 Comparing barcode and RFID technologies

Both barcode and RFID technologies are used widely for automatic identification and tracking, and each fulfills this function efficiently and reliably. However, there exist significant differences between these two technologies:

- Barcode technology works by scanning a printed barcode label with barcode readers such as an optical laser, while RFID scans a tag using radio frequency signals. Barcodes do not work well in environments where barcode labels are prone to being wrinkled, dirty or smudged. These are common problems because the barcode labels have to be exposed on the outside of the product. However, RFID technology can perform well in these environments because RFID tags can be implanted within the product itself. This means the electronic components of RFID tags are better protected and can therefore offer greater ruggedness and reusability.
- Barcode readers require labels to be in their optical line-of-sight, something that is not required of RFID readers. This makes reading RFID barcodes much less time-consuming, as the scanners do not have to be individually positioned to scan each label in turn; rather an RFID reader can read tens of RFID tags in its scanning area simultaneously, speeding up the process considerably.
- Barcode labels can only be read at a limited distance, typically no more than 15 ft. In contrast, RFID scanners can read tags at up to 300 ft.
- It is impossible to alter the information in a barcode label once it is printed out. However, the RFID reader can communicate with RFID tags and alter the information in each rewritable tag. In addition, an RFID tag can store much more information than a barcode.
- Barcodes can be replicated or counterfeited with ease, while RFID tags are much more secure due to their greater complexity.

Although this comparison primarily focuses on the advantages of RFID over barcodes, the use of both technologies is not mutually exclusive. It is not expected that one will simply replace the other, but rather both may work in tandem in different marketing areas. In some respects, barcodes have advantages over RFID. For example, compared with RFID, barcode technology is easier to use and more affordable.

1.4 RFID technology

To fully utilize RFID technology in apparel supply chain operations, practitioners must understand the fundamentals of RFID technology clearly. To this end, the

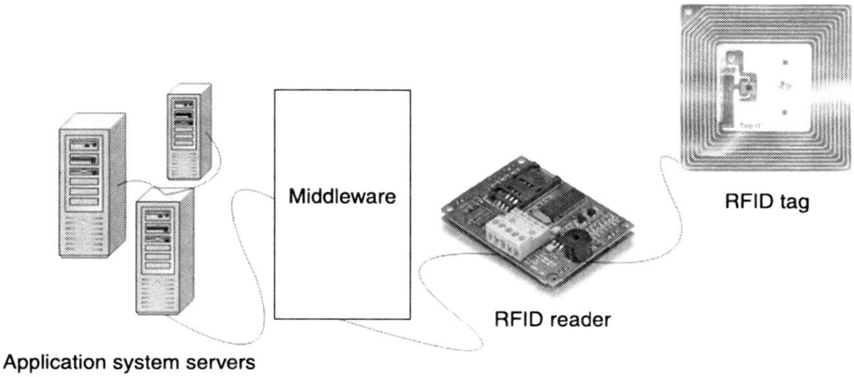

1.1 Architecture of an RFID system.

following section will outline the basics in order that the potential benefits of the technology may be fully understood. An RFID system usually consists of the three components shown in Fig. 1.1, which include:

1. RFID tags;
2. RFID readers;
3. middleware.

The RFID tag is attached to the item that is to be identified or tracked within the supply chain by the RFID system. The RFID reader provides multiple functions such as:

- powering and identifying RFID tags;
- reading (writing) data from (TO) tags; and
- communicating with the middleware.

The middleware can receive data from the RFID reader, and enter it into a database, as well as provide access to it. The communication mechanism between tag and reader is essential to the development of RFID systems.

1.4.1 RFID tags

RFID tags are the heart of the RFID system, which stores the information that describes the item being tracked. They come in a variety of shapes, sizes and capabilities. In general, an RFID tag is made up of three essential components, including:

1. *Antenna*: An antenna is responsible for transmitting and receiving radio waves between RFID tags and readers. It is an electrical device that converts electric currents into radio waves, and vice versa. In certain situations, an antenna can

collect enough radio waves to power the tag's other components without a battery. There exist various antennas with different shapes and sizes in the case of RFID.

2. *Integrated circuit*: The integrated circuit (IC) component is the brains of RFID tag. It is actually a simple microprocessor. For many tags, the purpose of IC is to transmit the tag's unique identifier. However, the IC also acts as the main controller if the tag includes some peripheral components, which is responsible for collecting any extra information and transmitting it along with the tag's identifier. Some ICs can store as much as 2 kilobytes of data. In addition, the IC is responsible for ensuring that tags do not transmit their information at the wrong time by using an appropriate transmission algorithm, which ensures that the tag transmits either in the proper time slot or at random intervals.
3. *Printed circuit board*: The printed circuit board is the material that holds the tag together. This can exist in various shapes and sizes as well as different types of materials including glass tubes, molded styrene or epoxy resin, etc. For example, tags that are used for tracking fabrics on a shop floor, where extremely high temperatures and humidity may be encountered, would tend to be very rigid and placed inside a protective enclosure. In contrast, RFID tags which track fabrics in warehouses, where such conditions are less extreme, require greater flexibility so that they can bend with the fabric. It should be pointed out that printed circuit boards cannot be made from some types of material due to their adverse effect on the readability of RFID tags. These include metal, liquid and other particularly dense materials.

RFID tags can be classified into two categories: passive tags and active tags. Passive tags do not have a dedicated power supply. Their operating powers are derived from the electrical field generated by the RFID reader. Consequently, passive tags cannot work unless they are placed near to the reader (usually within a few centimeters). RFID tags are usually very small and can be installed unobtrusively because they do not contain an internal power supply. Some are as light, or even lighter, than barcoded labels.

In contrast, active tags have their own power supply inside, usually in the form of a battery. They have significantly greater read range than passive tags and unlike passive tags, can be read through materials that are usually impenetrable to radio waves. However, active tags are much greater in size as well as expensive due to their more complex circuitry.

The choice between a passive or an active tag is purely dependent upon the application for which it will be used. Passive tags are usually utilized in situations where tag cost is a main concern, such as for production data capture in apparel sewing lines, where the tag can be placed near to the reader for fuel dispensing, for example. Active tags would be used in cases where the cost, size and battery life-span of the tag are less important concerns, as well as when tags need to be used at a greater distance. An example of this may be in areas such as toll-paying and shipping container identification.

1.4.2 RFID readers

To retrieve the data stored on an RFID tag, an RFID reader is required. A typical RFID reader is a component that has one or more antennas that emit radio waves and receive signals back from the tag. The RFID reader is responsible for orchestrating the communication between itself and RFID tags in its read range, which decodes the data encoded in the tag's IC and passes the data to the middleware for further processing. RFID readers can be classified into the following four categories:

1. *Handheld reader*: The handheld reader combines the power of a mobile device with wireless networking and multi-protocol RFID capabilities, which acts like a handheld barcode scanner.
2. *Vehicle-mount reader*: This is a reader embedded into mobile devices such as trucks and forklifts, which has the convenience of hands-free scanning with little or no human intervention.
3. *Fixed reader*: This is mounted in a fixed position to read tags automatically as items pass by or near them, which can help reduce the communications burden on your network or servers.
4. *Hybrid reader*: Each of the three readers described above is dedicated to reading either passive or active tags. The hybrid reader is designed to be able to switch between both passive and active modes in order to read both types of tags.

RFID readers must be capable of detecting minute changes, through their antennas, in the electromagnetic field which they generate. This is the way in which the tags communicate with the reader. The RFID reader emits radio waves in ranges of anywhere from 1 inch to 100 ft or more, depending upon its power output and the radio frequency used. In passive mode, the RFID reader creates a radio frequency field when it is turned on. RFID tags are activated if they are detected by a reader. These tags do not require battery power and they draw their energy from the radio frequency field. In the active mode, the RFID reader switches to the read mode and interrogates active tags when they come within its range.

1.4.3 Tag and reader communication

The communication between RFID tags and readers usually happens on the following radio frequency bands:

- *Low frequency (LF)*: This comprises of electromagnetic waves which emit radio frequencies of between 30 and 300 KHz. These are generally used in passive tags and in short-range applications, such as cutting component identification in sewing assembly lines and anti-theft systems in automobiles. LF RFID systems typically operate between 125 and 134 KHz and between 140 and 148.5 KHz. These systems provide a shorter read range (<0.5 m or 1.5 ft) and slower read speed than the higher frequencies. However, compared to any of higher frequency systems, LF systems have the strongest performance in the reading of tags which are attached to objects with water or metal content.

- *High frequency* (*HF*): This consists of radio frequencies in the range of 3 to 30 MHz. These are widely used in smart card and smart label applications, such as baggage tracking or small product labeling. HF RFID systems typically operate at 13.56 MHz, which have a greater read range and higher read speed than LF systems. Their typical read ranges are less than 1 m (3 ft), and they have a diminished ability to read tags on objects with a higher water or metal content compared with LF systems.
- *Ultra high frequency* (*UHF*): This refers to radio frequencies in the range of 300 MHz to 3 GHz, which are primarily used in highway toll-collection applications. Typically UHF RFID systems operate between 868 and 928 MHz, such as 868 MHz in Europe and 915 MHz in North America. The UHF system's read range is up to 3 m (9.5 ft) and the data transfer rate is faster than HF systems. Its ability to read tags on objects with water or metal content is superior to both HF and LF systems.

In summary, higher frequency systems require higher power and are somewhat more expensive than their lower-frequency counterparts. They support greater read distances as well as higher data transmission rates, and are more orientation-sensitive. Higher-frequency systems are less capable of penetrating metallic surfaces but are less prone to noise interference than are lower-frequency systems.

1.4.4 Middleware

Middleware is a key component for managing the flow of information between tag readers and application systems. To quantify the data, RFID systems must first collect it in an application system (software). RFID middleware applies filtering and formatting to data captured by a reader, and provides this processed data to back-end application systems.

Broadly speaking, the middleware refers to software or devices that connect the RFID hardware and the RFID application systems. The general architecture of the middleware component is composed of four layers (Fig. 1.2):

1. *Reader interface*: This is the lowest layer of the middleware component, and is that which handles the interaction with the RFID hardware. It maintains the drivers of all the devices supported by the RFID system, and manages all hardware-related parameters such as reader protocol, air interface and host-side communication.
2. *Data processor and storage*: This layer is responsible for processing and storing the raw data coming from the readers. Examples of processing logic implemented by this layer include data filtering, aggregation and transformation. If necessary, this layer can handle the data level events associated with a specific application.

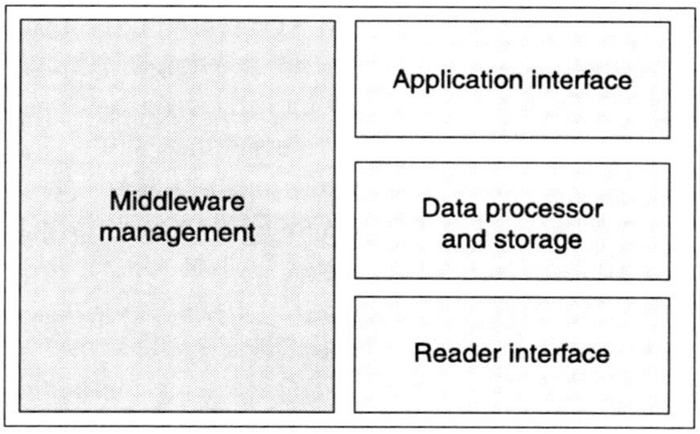

1.2 General architecture of a middleware component.

3. *Application interface*: The application interface layer provides the application with a program with which it can access, communicate and configure the middleware component. It integrates the application systems with the middleware by converting the requests from application systems to applicable middleware commands.
4. *Middleware management*: This layer is responsible for managing the configuration of the middleware component, which provides the following functions:

   - add, configure and modify connected RFID readers;
   - modify application level parameters such as filters, and duplicate removal timing windows;
   - add and remove services supported by the RFID middleware.

   By using the middleware component, the following advantages can be obtained:

   - The details of RFID hardware are hidden from the application systems.
   - Raw data from RFID readers can be processed before being passed as aggregated events to the application systems. Standardization is provided to handle the huge amount of raw data.
   - An effective method is provided to manage RFID readers and query the RFID data.
   - It can provide the user with an interface that allows the set-up and reconfiguration of the RFID devices.

1.5 RFID applications in the fashion supply chain

The apparel industry has been regarded generally as lagging behind other industries such as electronics and automotive industries in regards to the utilization of innovative technologies. However, with the emergence and popularity of RFID technology, a number of apparel enterprises seized the development opportunity of this technology quickly and utilized it to enhance their competitive edge. RFID technology has now been widely applied throughout the apparel industry, ranging from fashion retailers to department stores to logistics service providers (Moon and Ngai, 2008).

As an example, in the tracking of rental uniform items, in 2002 a German company invented two types of new washable RFID tags to target the apparel and uniform rental markets in the United States (Anon, 2002). In 2008, American Apparel implemented item-level RFID tagging and product tracking (O'Connor, 2008). Kaufhof and Marks & Spencer started their RFID activities in 2003 and in 2004, respectively, by using item-level RFID tags on individual items of clothing (Collins, 2006; Loebbecke and Huyskens, 2008). Christophe Cavailles, director of DHL Solutions Fashion, also stated that RFID was well suited to reducing the complexity of the inventory and delivery checking involved in the apparel supply chain (Ilic, 2004). Legnani *et al.* (2010) have pointed out that the apparel industry, characterized by short life-cycles, quick-response production, fast distribution, erratic customer demands and impulsive purchasing, is one of the sectors which can most extensively benefit from the RFID technology.

RFID technology has also been widely used to develop real-time production data capture systems. This is used whilst collecting production data from sewing assembly lines and other departments in apparel manufacturing (Guo, 2008). Guo *et al.* (2009) integrated the RFID-based data capture system with an artificial intelligent algorithm to provide effective production decisions for apparel manufacturers. Ngai *et al.* (2009) utilized a case study approach to explore the adoption of RFID-based information systems in apparel manufacturing, and identified various strategies for the successful application of the technology in apparel manufacturing. Choy *et al.* (2009) developed a sample management system by integrating RFID technology and a case-based reasoning technique. To manage the iterative process for evaluating fabric swatches in the development of new apparel products, RFID technology is utilized to provide real-time swatch tracking and accurate fabric status updates within a fabric sample storeroom.

Madhani (2011) pointed out that fashion retailing may greatly benefit from the application of RFID technology. He proposed a marketing value-added framework by using the item-level tagging of RFID in order to help fashion retailers achieve overall business excellence. Wong *et al.* (2006) employed RFID technology to assist businesses in checking the apparels' authenticity. This involved using passive RFID tags to identify cloned RFID tags. Moon and Ngai (2008) proposed a four-point plan for an RFID-based framework, which would improve the business values of fashion retailers. In this framework, RFID-based decision-making solutions create business values for fashion retailers through improving responsiveness, relatedness and refinement. Hwang and Rho (2008) proposed an RFID system for centralized reverse SCM. This employed the RFID-based smart shelf/fitting room and inbound/outbound facilities to collect data and generate useful information, such as attractiveness of product and product return type. Al-Kassab (2010) examined the business value of RFID technology in apparel retailing by conducting a case study at Galeria Kaufhof, one of the largest department store chains in Europe, and pointed out RFID technology's usefulness in discovering process inefficiencies, cutting costs and improving the quality of service.

Some researchers have demonstrated the useful application of RFID technology in other aspects of the apparel supply chain, with the exception of apparel retailing and manufacturing. Wong *et al.* (2006) investigated the information flow by means of RFID-based data throughout the apparel supply chain. For this, he used an entity relation diagram to observe the relationship between each of its members in the supply chain. Bottani *et al.* (2009) examined the impact of RFID technology on logistical processes in the fashion retail supply chain by using quantitative assessments. This study concluded that RFID technology has the potential to generate improved revenues for the fashion supply chain.

1.6 References

Al-Kassab, J., Thiesse, F. and Fleisch, E. (2010), RFID in the apparel retail industry: a case study from Galeria Kaufhof. In: Damith, C. R., Quan, Z. and Shang, S. Z. (eds), *Unique Radio Innovation for the 21st Century: Building Scalable and Global RFID Networks*, London, New York, Heidelberg: Springer Dordrecht, 281–308.

Anon (2002), New direct-to-textile washable tag, *RFID Journal*, 6 August 2012.

Bottani, E., Ferrettia, G., Montanaria, R. and Rizzia, A. (2009), The impact of RFID technology on logistics processes of the fashion industry supply chain, *International Journal of RF Technologies: Research and Applications*, 1(4), 225–52.

Choy, K. L., Chowa, K. H., Moon, K. L. and Zeng, X. (2009), A RFID-case-based sample management system for fashion product development, *Engineering Applications of Artificial Intelligence*, 22, 882–96.

Collins, J. (2006), Marks & Spencer to extend trial to 53 stores. *RFID Journal*.

Guo, Z. X. (2008), Intelligent production control decision-making for apparel manufacturing process. In: *Institute of Textiles and Clothing*, Hong Kong: The Hong Kong Polytechnic University.

Guo, Z. X., Wong, W. K., Leung, S. Y. S. and Fan, J. T. (2009), Intelligent production control decision support system for flexible assembly lines, *Expert Systems with Applications*, 36(3), 4268–77.

Hwang, Y.-M. and Rho, J.-J. (2008), RFID System for centralized reverse supply chain in the apparel industry, *The Second International Symposium and Workshop on Global Supply Chain, Intermodal Transportation and Logistics,* Korea.

Ilic, C. (2004), Tracking fashion with RFID, *RFID Journal*.

Legnani, E., Cavalieri, S., Pinto, R. and Dotti, S. (2010), The potential of RFID technology in the textile and clothing industry: opportunities, requirements and challenges. In: Damith, C. R., Quan, Z. and Sheng, S. Z. (eds), *Unique Radio Innovation for the 21st Century: Building Scalable and Global RFID Networks*, London, New York, Heidelberg: Springer Dordrecht, 309–29.

Loebbecke, C. and Huyskens, C. (2008), A competitive perspective on standard-making: Kaufhof's RFID project in fashion retailing, *Electronic Markets*, 18(1), 30–8.

Madhani, P. M. (2011), RFID deployment: fast fashion retailing, *SCMS Journal of Ind. Management*, 8(2), 40–51.

Moon, K. L. and Ngai, E. W. T. (2008), The adoption of RFID in fashion retailing: a business value-added framework, *Industrial Management and Data Systems*, 108(5), 596–612.

Ngai, E. W. T., Moon, K. L., Chain, A., Chan, B. and Wu, W. (2009), A case analysis of adoption of an RFID-based garment manufacturing information system, *Proceedings of the 2009 Pacific Asia Conference on Information Systems (PACIS)*, 32.

O'Connor, M. C. (2008), American Apparel makes a bold fashion statement with RFID, *RFID Journal*.

Shepard, S. (2005), *RFID: Radio Frequency Identification*, New York: McGraw-Hill.

Wong, K. H. M., Chan, A. C. K., Hui, P. C. L. and Patel, C. A (2006), A framework for data flow in apparel supply chain using RFID technology, *Industrial Informatics, 2006 IEEE International Conference*.

Wong, K. H. M., Hui, P. C. L. and Chan, A. C. K. (2006), Cryptography and authentication on RFID passive tags for apparel products, *Computers in Industry*, 57, 342–9.

2

The role of radio frequency identification (RFID) technologies in improving distribution and retail operations in the fashion supply chain

E. BOTTANI, A. VOLPI, A. RIZZI, R. MONTANARI and M. BERTOLINI, University of Parma, Italy

DOI: 10.1533/9780857098115.13

Abstract: This chapter examines the benefits achievable when radio frequency identification (RFID) technology is implemented into the fashion supply chain, and is based on two of our earlier studies in this field. A theoretical feasibility study demonstrates how the logistics processes of a typical fashion retail store (RS) and distribution centre (DC) should be reengineered to implement RFID technology successfully, and provides a quantitative evaluation of the impact of applying RFID technology to those processes. These preliminary findings are confirmed by the main outcomes obtained from an in-field implementation, the RFID Fashion Pilot (RFP) study. The results of our economic analysis could be useful for companies operating in the RFID area or in fashion, and help to provide an economic justification for implementing RFID in the fashion industry.

Key words: radio frequency identification (RFID), fashion supply chain, feasibility study, pilot study.

2.1 Introduction

The fashion industry is becoming increasingly complex and dynamic, partly due to the advent of own brands and the globalization of sourcing and supply chain decisions (Hines and McGowan, 2005). The industry has recently attracted the attention of several researchers in the area of operations and supply chain management (SCM) (Bruce *et al.*, 2004; Christopher and Towill, 2001; Harrison *et al.*, 1999; Lowson *et al.*, 1999). The fashion supply chain has specific characteristics which make its management particularly complex, including short product life-cycles, high volatility and low predictability of demand (Bruce *et al.*, 2004; Choi, 2007; Christopher *et al.*, 2004).

Balancing production and demand is a challenge faced by all manufacturing businesses, including fashion retailers (Nair and Closs, 2006). Such industries have learned that the supply chain operates more efficiently when lead times are decreased and forecasting errors are reduced. Supply chain performance can be improved by identifying approaches and tools for automating logistics processes (thus reducing the time to market), and improving the availability of information (Choy *et al.*, 2009).

The fashion industry has particular challenges because demand is highly variable, influenced by rapidly changing trends, and the selling season is very short (Brun *et al*., 2008). The fashion industry also produces a wide range of complex textile products, including casual wear, outdoor clothing, intimate apparel, fashion and interior textiles, and specialist clothing such as sportswear and work wear (Lo *et al*., 2008). Recently, the advent of 'fast fashion' has exacerbated the problem for the industry. Characterized by shorter life-cycles, quicker production responses, faster distribution, more erratic customer preferences and impulsive purchasing, this business trend places significant pressure on each segment along the supply chain and creates tremendous challenges for fashion enterprises. To maintain competitiveness in the global market, fashion manufacturers need to become faster at delivering new products to clients and responding to market changes (Jacobs, 2006).

RFID technology could improve coordination between marketing and manufacturing and thus offers an attractive way to improve supply chain efficiency (Taylor *et al*., 2003; Zahay and Handfield, 2004). In particular, RFID enables real-time tracking of items, safe monitoring and automation of warehouse operations, which can lead to increases in sales volumes and improved profitability for suppliers and retailers (Böse and Pal, 2005; Loebbecke, 2005). Applying RFID across a business organization has a significant impact on the entire supply chain, from suppliers to retailers (Evans and Berman, 2001; Roy *et al*., 2004).

In the context of fashion, RFID technology has the potential to help retailers solve several key issues. First, since the product life-cycle of fashion items is very short, it is vital to ensure that the product is delivered to the store as quickly as possible. Since RFID provides the capability to track and trace items, it could streamline many activities along the supply chain. Second, fashion retailers often need space to display products of different styles, sizes and colours; consequently, managing items efficiently at retail stores (RSs) is important. An RFID-based stock management system can help employees to identify items needed by customers more quickly and accurately (Reynolds *et al*., 2006). Third, RFID gives each clothing item a special identifier, which can help towards preventing or at least reducing the sale of counterfeit and illegal products (Moon and Ngai, 2008). Finally, RFID tags could replace acousto-magnetic tags or RF tags for electronic article surveillance (EAS) applications (*RFID Journal*, 14 October 2009).

The following discussion is based on three studies we carried out on RFID application in the fashion supply chain. In the next section, we provide a preliminary, theoretical analysis of the economic case for implementing RFID in the fashion context. Using a feasibility study, we show how the logistics processes of a typical RS and DC for fashion items should be reengineered to implement RFID technology successfully, and provide a quantitative evaluation of the impact of RFID technology applied to those processes. Section 2.3 describes the main outcomes obtained from an in-field implementation, the RFID Fashion Pilot (RFP), which was conducted to confirm the theoretical findings. The findings are useful for companies operating in the RFID field itself and in the fashion industry, and provide an economic justification for adopting RFID technology in the fashion industry.

2.2 Assessing the feasibility of using RFID technology: process reengineering

In this section, we describe how the logistics processes of a typical fashion RS and DC can be reengineered to implement RFID technology successfully, and discuss the related benefits.

2.2.1 Research methodology

The results presented in this section are based on the direct examination of 11 sites belonging to companies operating in the fashion industry, including 6 DCs and 5 RSs, in a 2-echelon supply chain, and on a detailed mapping of the related processes.

Adopting the business process reengineering (BPR) approach (Hammer and Champy, 1993), the current (AS IS) supply chain processes and related performance for each site were examined. Reengineered (TO BE) scenarios were subsequently developed, hypothesizing the adoption of RFID technology and of the Electronic Product Code (EPC) system for process automation. A detailed feasibility study was performed based on the gap between current and reengineered processes (AS IS vs. TO BE).

The supply chain analyzed was defined by a panel of experts, which included 15 members from 13 fashion companies representing both Italian and international enterprises, among them Aeffe, Branded Apparel, Dolce & Gabbana Industria, Imax, Lotto Sport Italia, Miroglio and Trussardi, all of which participate in the research activities of the RFID Lab of the University of Parma. The panel members were selected from people reporting directly to the firms' top management and covering the key business functions where RFID technology has the potential to make a significant impact, including logistics, procurement, operations and information. Panel members were involved in monthly round-table discussions, which took place at the Department of Industrial Engineering of the University of Parma between June and December 2008.

During the preliminary meetings, panel members were asked to indicate which sections of the fashion supply chain should be evaluated for the potential impact of RFID technology. It was agreed that the analysis should start at the end of the production lines at the manufacturer's site, and should include a DC and one or more RSs. Logistics processes at the manufacturer's site were not considered to assess the impact of RFID in this study. The choice was motivated by the fact that it would be difficult to apply tags to clothes at the manufacturer's site, as industrial processes such as washing and ironing may be performed at this stage.

Panel members were also involved in the selection of case study sites where the internal processes could be analyzed. Seven of the companies involved agreed to be examined through site visits and direct interviews, and were invited to indicate a DC and/or a RS that would be illustrative of the company's internal processes and suitable for direct investigation. DCs and RSs with a large flow of items, highly automated logistics processes or that had adopted advanced information systems, were preferred. Eventually, the 11 sites were identified and investigated, 6 DCs and 5 RSs. All the sites were located in Italy. Site visits and the data collection phase lasted from July to November 2008. The relevant characteristics of the sites investigated and a list of visits performed are indicated in Table 2.1 (for the purposes of confidentiality, the names of the companies involved in the research have been omitted).

The data collection phase involved questionnaires and personal interviews with the managers of the chosen sites. Two questionnaires were designed, to collect quantitative and qualitative information concerning the macro-processes characterizing the activities of the DCs and RSs, the processes examined having

Table 2.1 Summary of visits performed for data collection and main characteristics of the companies examined

Company	Relevant information	DC	RS	Date
Company 1	Items handled: footwear, sportswear and accessories RS type: monobrand Supply chain configuration: 1 DC, which serves about 20 direct operating stores	X	X	14 July 2008
Company 2	Items handled: underwear RS type: monobrand Supply chain configuration: 2 DCs, serving 7 outlets and 100 RSs (including 5 direct operating stores)	X	X	15 July 2008 8 October 2008
Company 3	Items handled: women's wear and accessories RS type: monobrand Supply chain configuration: 2 DCs, serving about 130 RSs	X		6 November 2008
Company 4	Items handled: women's and men's wear and accessories RS type: monobrand Supply chain configuration: 1 DC, serving about 40 direct operating stores	X	X	3 October 2008
Company 5	Items handled: women's and men's wear, underwear, beachwear, accessories RS type: monobrand Supply chain configuration: 1 DC, serving 14 department stores		X	6 October 2008

Table 2.1 Continued

Company	Relevant information	DC	RS	Date
Company 6	Items handled: women's and men's luxury wear, underwear, beachwear, accessories, jewelry RS type: monobrand Supply chain configuration: 1 DC, serving 17 direct operating stores	X		24 September 2008
Company 7	Items handled: women's and men's luxury wear, underwear, accessories, perfumes, jewelry RS type: monobrand Supply chain configuration: 1 DC, serving 19 direct operating stores	X	X	11 November 2008

been suggested by the panel of experts. Specifically for the DCs, the processes investigated through the questionnaire included:

1. receiving;
2. storage;
3. inventory management;
4. picking and sorting;
5. stock movement;
6. packing and marking;
7. shipping; and
8. return flows (from RSs or to manufacturer).

The questionnaire designed for the RS encompassed:

1. receiving;
2. inventory management/replenishment;
3. re-pricing;
4. check-out; and
5. return flows (from customers or to DCs) and shipments to other RSs.

Particular attention was paid to activities that could be automated or improved by the adoption of RFID technology. In this regard, specific questions included in the questionnaires covered the following performance parameters:

- accuracy of the processes investigated (in terms of number of errors occurring in each process);
- labour efficiency (in terms of time required to complete the process); and
- stock level and inventory management.

The questionnaires were completed during visits to each site. A detailed list of the average values of the quantitative parameters of supply chain processes collected during the survey phase is presented in Table 2.2.

Table 2.2 Average quantitative data resulting from the data collection phase

Supply chain echelon	Quantitative data of supply chain processes	Measurement unit	Numerical value (average)
Distribution center	**Receiving**		
	Number of receiving gates	Gates	3
	Annual number of orders received	Orders/year	1500
	Annual number of items received	Items/year	7850000
	Annual number of cases received	Cases/year	350000
	Number of items per case	Items/case	22
	Number of cases per pallet	Case/pallet	20
	Storage		
	Number of employees in the storage area	Employees	8
	Number of storage locations	Locations	45000
	Inventory management		
	Number of employees devoted to inventory counts	Employees	29
	Annual number of inventory counts	Inventory/year	2
	Time required for inventory count	Hours/inventory	40
	Stock movements		
	Percentage of cases to be moved (out of the total number of items received)	%	2%
	Picking and sorting		
	Number of employees in the picking area	Employees	14
	Number of items moved during picking operations	Items/year	7850000
	Annual number of picking lists fulfilled	Picking lists/year	200000
	Packing and marking		
	Number of packing stations	Stations	6
	Shipping		
	Number of items per case shipped	Items/case	10

	Return flows (from RSs or to manufacturer)		
	Annual number of items returned from RSs	Items/year	160000
	Annual number of items returned to manufacturers	Items/year	400000
	Number of manufacturing processes required per item returned	Processes	2
Retail store	**Receiving**		
	Number of employees in the receiving area	Employees	2
	Annual number of items received	Items/year	94000
	Inventory management		
	Number of employees for inventory counts	Employees	5
	Annual number of inventory counts	Inventory/year	2
	Time required for inventory count	Hour/inventory	24
	Number of doors connecting to the expositive area	Doors	1
	Re-pricing		
	Annual number of re-pricing per item	Re-pricing	6
	Number of items available in the expositive area	Items	10000
	Check-out		
	Number of cash registers	Cash registers	2
	Annual number of items sold	Items/year	89000
	Return flows (from customers or to DCs/RSs) or shipments to other RSs		
	Annual number of items returned from customers or sent to DCs or shipped to other RSs	Items/year	5000

2.2.2 AS IS analysis

The AS IS analysis examined the manufacturing and logistics processes that characterized the supply chain echelons investigated. Specifically, the aim of the analysis was to describe all of the manufacturing and logistics activities that clothing items undergo in the fashion supply chain. The results of the AS IS analysis were discussed and validated through periodic meetings with the panel of experts.

As a result of the AS IS analysis, a detailed examination of the logistics processes was derived, and it emerged that, in the case study companies analyzed, DC and RS processes could be managed in several different ways (Table 2.3). For instance, there were five possible procedures for receiving operations at the DC, ranging from manual reading of each clothing item (AS IS-1) up to fully automated reading of items handled by a sorter (AS IS-4). Depending on the AS IS scenario, the impact of RFID implementation could be different. In the case of AS IS-1, implementing RFID technology in the receiving process has the potential to substantially improve the performance of receiving operations, by removing manual reads and avoiding (or at least reducing) possible related errors; conversely where the receiving procedures are fully automated or performed efficiently (e.g. AS IS-4), introducing RFID technology would have a limited impact. Moreover, in the AS IS-5, receiving operations are outsourced; it is thus reasonable that, even if RFID is implemented in the DC, the process will still be outsourced and so there is no benefit from implementing RFID for that process.

2.2.3 TO BE reengineering

Using the results of the AS IS analysis, the TO BE reengineering developed new scenarios where RFID and EPC might optimize the processes examined. TO BE scenarios were developed to ensure:

- traceability of clothing items, and
- efficient management of manufacturing and logistics processes.

In order to be representative of real processes, the TO BE scenarios were discussed with the panel of experts, with the aim of validating the information, and operational and technological parameters of the scenarios developed. The parameters were defined for each process examined during the AS IS analysis.

The information parameters concerned pieces of information to be stored in the tag chip or EPCIS (EPC Information Services) repository (EPCglobal, 2004, 2007), as well as data to be shared between logistics processes to optimize management. The operational parameters of the TO BE scenarios encompassed the definitions of new procedures for employees, according to the BPR methodology. The technological parameters concerned the hardware and software infrastructure hypothesized for RFID deployment. Defining the hardware infrastructure required identifying tags, readers and other equipment to be used in the TO BE processes. The panel of experts suggested considering item-level tagging when designing the

Table 2.3 Summary of supply chain processes resulting from the AS IS analysis and corresponding TO BE reengineering

	Process examined	Possible AS IS scenarios	TO BE scenarios
DC	Receiving	AS IS-1: items received are identified by reading their barcodes; a contextual check of quantity and error is performed by employees. AS IS-2: data related to cases received are manually inserted in the company's IS. Cases are then barcode labeled and read. Quantity checks are limited to a sample of cases, for which all items contained are manually counted. Error checking consists of opening all cases received and verifying the correspondence between items received and ordered. AS IS-3: cases received are identified by reading their barcodes; quantity and errors checking is analogous to AS IS-2. AS IS-4: identification is automated by reading barcodes of items handled by a sorter; a contextual check of quantity and error is performed by employees. AS IS-5: the process is outsourced.	Each receiving dock is equipped with an RFID handheld device, which is used to read RFID tags of cases or items received. Thanks to the EPC Network, information on items or cases received is available in advance to the DC; hence, when reading RFID tags of items/cases received, quantity and mix checks can be real-time performed. If a misalignment is identified on some cases or items received, the employee performs a further manual check, by reading barcodes of those cases or items. An RFID printer is required in case labels should be reprinted or modified.
	Storage	AS IS-1: barcodes of cases are read to obtain, from the company's IS, the information concerning the storage position. Then barcodes of the storage location and of the case are read to confirm the storage operation. AS IS-2: barcodes of cases are read to obtain, from the company's IS, the information concerning the storage location. Storage operations are not confirmed. AS IS-3: the employee identifies a suitable storage location for the case. Once the case is stored, the corresponding data are manually inserted in the company's IS. AS IS-4: the location information and the confirmation of storage are automatically managed by the company's IS. This is typically the case of an AS/RS warehouse.	Each employee is provided with an RFID wearable device. Moreover, each storage location will be labeled with RFID tags, in order to be uniquely identified. By reading RFID tags of cases, the employee automatically obtains the information concerning the warehouse location when the product should be stored. Upon storage, the employee reads the RFID label of location and cases, to store the corresponding information in the company's IS.

(*Continued*)

Table 2.3 Continued

Process examined	Possible AS IS scenarios	TO BE scenarios
Inventory management	AS IS-1: during inventory counts, the employee reads the barcodes of items, packages and their locations. AS IS-2: the inventory count is automated, being supported by an AS/RS warehouse that handles items close to the employee, which reads the barcodes. AS IS-3 (only for inventory counts concerning a specific item): the process is not performed.	Each employee involved in inventory counts is provided with an RFID 'totem', thus automating RFID reads of items, cases and storage locations. Due to real-time connection with the company's IS, during counting it is possible to directly check the correspondence between RFID reads and information available in the IS. If a lack of correspondence occurs, the employee performs a further manual check.
Picking and sorting	AS IS-1 (batch picking): the employee follows a picking list and confirms the picking operation by ticking elements on the list. AS IS-2 (batch picking): an AS/RS warehouse moves items to a location, where an employee reads barcodes of items or cases picked. AS IS-3 (order picking): picking is performed following a 'Look&Pick' mechanism; the confirmation is automated. AS IS-4 (order picking): during picking, the employee reads the barcode of cases, items and of the stock keeping unit (SKU) being prepared. AS IS-5 (order picking): the employee follows a picking list and confirms the picking operation by ticking elements on the list.	Each employees operating in the picking area is provided with an RFID wearable device. A picking list is displayed on the RFID device; during picking, RFID tags of products and picking locations are read. Due to RF connection, the company's IS is real-time updated.
Stock movements	AS IS-1: any time that cases or items are moved to a different location in the warehouse, the corresponding barcodes are read by an employee.	RFID wearable devices are used to read cases of products and storage locations during stock movements. RFID tag of the new location is read to update the company's IS.

Packing and marking	AS IS-1: when creating the SKU for shipping, the employee reads the barcode of items picked, cases and of the SKU being prepared. AS IS-2: the employee reads the barcode of picking list, cases and of the items picked. AS IS-3: the employee reads the barcode of each item picked. AS IS-4: the employee puts items picked in a new SKU. As items are moved by a sorter, they are automatically read. A manual checking is performed on a sample of SKUs, by reading the barcodes of items contained. AS IS-5: cases are prepared for shipping during picking operations. Hence, packing and marking are not performed.	When items arrive at the picking station, the employee reads the corresponding RFID tag and the tag of the SKU being packaged. Each packing station is also equipped with an RFID printer, to create labels for cases to be shipped, and an RFID gate, to read the corresponding label and identify the content.
Shipping	AS IS-1: to check the content of SKUs prepared for shipping, the employee reads the barcodes of cases; the company's IS is real-time updated through RF connection. AS IS 2: the employee reads barcodes of cases and of the shipping location; the company's IS is real-time updated through RF connection. AS IS 3: a shipping list is printed. The employee checks the correspondence between the list and the SKUs prepared for shipping by ticking cases to be shipped on the list. AS IS 4: the process is not performed.	Each shipping bay is equipped with an RFID gate, where cases to be shipped are passed through for reading.
Return flows (from RSs or to manufacturer)	AS IS-1: the employee reads the barcodes of all items returned. AS IS-2: barcodes of items returned are automatically read by a sorter. AS IS-3: items returned are not identified.	Returned flows are treated as received ones, that is RFID tags of each item or case are read and the corresponding information is real-time stored in the company's IS. RFID handheld devices are used to this extent. The required documents are automatically printed by an RFID printer.

(*Continued*)

Table 2.3 Continued

	Process examined	Possible AS IS scenarios	TO BE scenarios
RS	Receiving	AS IS-1: barcodes of items received are manually read; then the price is applied on the item's label. Usually, pricing is performed prior to exposure of items on the store shelves. AS IS-2: barcodes of items received are manually read; pricing is not performed, as the label already contains the price of product. AS IS-3: the employee reads barcodes of cases received and manually counts items contained in a sample of cases. AS IS-4: items received are manually counted; contextually, the price is automatically printed and applied on the item's label.	In the case of AS IS-1 and AS IS-4, it is assumed that each employee is provided with a wearable RFID device. Employees pick items received, and automatically read their RFID tags. An RFID printer is used to print the item's label, which is then attached to the item. In the case of AS IS-2 and AS IS-3, employees are provided with handheld RFID devices, which are used to read RFID tags of cases and items contained. By means of RFID reads, employees also automatically check the correspondence between items received and those expected. A further manual check is performed in the case a lack of correspondence occurs.
	Inventory management	AS IS-1: during inventory counts, barcodes of items are manually read by an employee. AS IS-2: during inventory counts, barcodes of items and cases are manually read by an employee.	Employees are provided with RFID handheld devices, which are used to read the RFID tags of items during inventory counts.
	Re-pricing	AS IS-1: barcodes of items are read by an employee; the discounted price of product is automatically printed and applied on the item's label. AS IS-2: barcodes of items are read by an employee; the new price is manually written by the employee on the item's label.	Employees are provided with RFID wearable devices, which are used to read the tags of items in the retail store. An RFID printer, connected with the RFID devices, automatically prints the new price to be attached to items.

Check-out	AS IS-1: barcodes of items sold are read during check-out; the company's IS is real-time updated. AS IS-2: barcodes of items sold are read during check-out; the company's IS is real-time updated. Antitheft devices are removed from the items.	Cash registers of the retail store are replaced with check-out points, which automatically read tags of items during check-out operations. By means of RFID reads, the stock available at the retail store is real-time updated. When required, the employee removes the antitheft device from items.
Return flows (from customers or to DCs/RSs) or shipments to other RSs	AS IS-1: items to be returned or shipped are manually read and an items list is automatically generated by a printer. The company's IS is real-time updated due to RF connection. AS IS-2: cases to be returned or shipped are manually read by an employee. The company's IS is real-time updated due to RF connection.	Employees are provided with RFID wearable devices, which are used to read the tags of items/cases to be returned or shipped to other retail stores. By means of RFID reads, shipping documents are automatically generated and printed.

TO BE scenarios, as well as equipping cases and pallets (when used) with RFID tags. It was also recommended to add RFID tags to the existing barcodes of items, cases or pallets. Technological solutions were tested in the laboratory to assess their feasibility and the achievable reading performance. The resulting TO BE processes are detailed in Table 2.3.

By comparing the AS IS processes with the reengineered ones, it is immediately clear that considerable benefits can be achieved in improved automation, labour efficiency and accuracy of operations using RFID. For instance, deploying RFID at the item level in the receiving process, by reading products received through an RFID handheld device, could make this process fully automated. It also allows the correspondence between the products received and those expected to be checked, as information on items or cases received is available in advance on the EPCIS infrastructure. When a lack of correspondence occurs, the employee performs a further manual check, by reading the barcodes of cases or items. The reengineered process reduces the number of manual operations required in several receiving process AS IS scenarios (i.e. AS IS-1 to AS IS-3); conversely, as mentioned previously, no substantial benefits emerge from deploying RFID in the case of AS IS-4 and AS IS-5.

2.2.4 Cost-benefit analysis

Based on the gap between the AS IS and TO BE scenarios, a detailed analysis of costs arising (e.g. cost of manpower for additional operations required) and savings resulting from RFID implementation (e.g. removal of errors or avoidance of manual operations) was performed for each DC and RS process investigated, producing a cost/saving balance for each one. The assessment was supported by laboratory experiments, which allowed an estimate of the time required to perform both manual operations and RFID-enabled procedures on items to be derived. Experiments in the RFID lab of the University of Parma were designed to replicate such activities at full scale.

During the economic assessment, the average values of the quantitative parameters collected during the site visits and reported in Table 2.2 were used, so as to derive general results that could be considered representative of the whole fashion supply chain. The cost of investments in RFID hardware and software in the TO BE processes was also considered, with costs of equipment being obtained from available sources, namely literature, market prices, catalogues of producers and vendors of RFID equipment. The unitary tag cost was set at 0.08 €/tag, commensurate with the large number of tags purchased.

Costs and savings resulting from implementing RFID may vary depending on the specific AS IS scenario of a particular process, which means that it is not possible to provide general cost/saving outcomes for DC and RS processes. In our theoretical model, costs and savings were computed for each process as a function of the AS IS scenario. As Table 2.3 indicates, the total number of AS IS scenarios is 30 for DC and 14 for RS; such scenarios can also be combined, obtaining about 18 000 configurations for DC and 128 for RS. From a mathematical point of view, this eventually leads to 2 304 000 possible supply chain configurations. During the elaboration of the results, this number was reduced to about 1 873 920, obtained by unifying AS IS scenarios where the RFID technology has the same impact (e.g. AS IS-4 and AS IS-5 of receiving processes at the DC), and by excluding 'unfeasible' configurations from the analysis. For instance, AS IS-4 of storage at the DC suggests the presence of an AS/RS (automated storage and retrieval system) warehouse; this could only be combined with a similar warehouse structure in the remaining processes (e.g. AS IS-2 of picking and sorting), while other combinations are very unlikely to be found in real cases.

The economic analysis was thus performed on the resulting 1 873 920 supply chain configurations. Among them, two specific DC and RS scenarios were examined in greater detail, identified as the 'best' and the 'worst' scenarios. They respectively indicated the AS IS supply chain configurations generating the highest and the lowest cost/saving balance following RFID implementation. Investments and cost/saving balances for the 'best' and 'worst' configurations are given in Table 2.4, to provide an overview of the boundary conditions. From the table, it can be seen that some processes exhibit a negative cost/saving balance under the worst configuration; this reflects the fact that, compared with the AS IS scenario, the implementation of RFID technology requires additional costs to manage the process, while benefits are limited or nil.

We can also see from Table 2.4 that additional processes, namely 'tagging' and 'other costs/savings', were considered in the economic analysis. The former encompasses the costs for tagging items, which involves investments in both tags purchased and the required manpower. Such costs are charged in full to the DC, while they are not considered for the RS, as it is assumed that when products are received by the RSs, they are already equipped with RFID tags. For the DC, other costs/savings mainly include costs for EPC global subscription, software integration and maintenance of the RFID system. The amounts were estimated based on commercial sources available. For RSs, besides the above cost components, other costs/savings also include benefits resulting from increased sales volume (and thus, increased turnover) as a consequence of RFID deployment.

Findings from a recent pilot project involving 280 American Apparel fashion stores showed that the implementation of RFID has the potential to improve sales volume of RSs by almost 15% (*RFID Journal*, 12 December 2008). However, this study lacks some information crucial to using this figure correctly; for instance, no information is available concerning the kind of shop investigated during the pilot study, or the factors causing the sale increase. Hence, trying to provide an exhaustive analysis of the turnover increase at the RS, we have examined four possible values, namely:

1. 0%, corresponding to no increase in turnover. Under some circumstances, the RFID could have no impact on the sales volume of the RS, as this depends on

Table 2.4 Cost/saving balance

		Investment [€]	Costs/saving balance [€/year]	
			'best' configuration	'worst' configuration
Distribution center	Tagging	7000.00	−656000.00	−656000.00
	Receiving	11000.00	107849.20	–
	Storage	35600.00	18958.33	−1266.67
	Inventory	30000.00	33170.00	500.00
	Stock movements	–	145.83	145.83
	Picking and sorting	56000.00	85041.67	–
	Packing and marking	110500.00	85090.83	673.14
	Shipping	18000.00	14583.33	–
	Return flows	–	13290.91	−727.27
	Other costs/savings	88800.00	−29000.00	−29,000.00
	Total	**356900.00**	**−326869.90**	**−685674.97**
Retail store	Receiving	11000.00	26111.11	865.32
	Inventory management	8500.00	5120.00	800.00
	Re-pricing	16000.00	17777.78	222.22
	Check-out	12000.00	2472.22	2472.22
	Return flows	–	58.33	13.89
	Other costs/savings	27611.00	16190.00	16190.00
	Total	**75111.00**	**67729.44**	**20563.65**

the specific AS IS scenario. For instance, if the inventory control at the RS is particularly efficient in the AS IS scenario, the turnover increase due to RFID would be limited.

2. 7.5%, corresponding to half the turnover increase observed in the pilot study available in the literature (referred to above).

3. and 4. 5% and 2.5%, corresponding to intermediate situations.

2.3 Assessing the feasibility of using RFID technology: profitability analysis

On the basis of cost/saving balance and investments, a detailed feasibility study was performed in order to evaluate the profitability of the RFID investment, in terms of net present value (NPV), payback period (PBP), internal rate of return (IRR) and return on investment (ROI). For computational purposes, a 5-year period and a 5% interest rate were assumed. The following selected results were derived from the economic analysis.

2.3.1 NPV of investing in RFID for the RS

The profitability for a fashion RS investing in RFID has been computed, considering all possible AS IS configurations and assuming the above-mentioned values of turnover increase. Sixty-four AS IS configurations were examined, resulting from combinations of the AS IS scenarios.

Figure 2.1 shows the NPV investing in RFID for an RS, as a function of the AS IS configurations examined and of the hypothesized value of turnover increase.

The NPV of the RFID investment exhibits a linear trend as a function of the hypothesized percentage of turnover increase, as demonstrated by the fact that the effect of different values of this parameter translates into a rise in the NPV curve. For a turnover increase of 0%, the profitability of the RFID investment depends on the AS IS configurations of the RS. Clearly, the investment is profitable (~€80 000) under the 'best' RS configuration, and some 30 additional AS IS configurations exhibit a positive (or non-negative) NPV for the investment. Conversely, the RFID investment generates negative NPV under the remaining AS IS configurations.

For a turnover increase of 2.5% or higher, all AS IS configurations of the RS show a positive NPV for the RFID investment over the 5-year period. The maximum NPV accounts for about €250 000, and is obviously achieved under the best RS configuration with a turnover increase of 7.5%. Under this scenario, the NPV ranges from approximately €250 000 to €120 000 as a function of the AS IS configuration considered. The profitability of the investment decreases from €140 000 (under the best RS configuration) to about €1 700 (under the worst RS configuration) when the turnover increase is set at 2.5%.

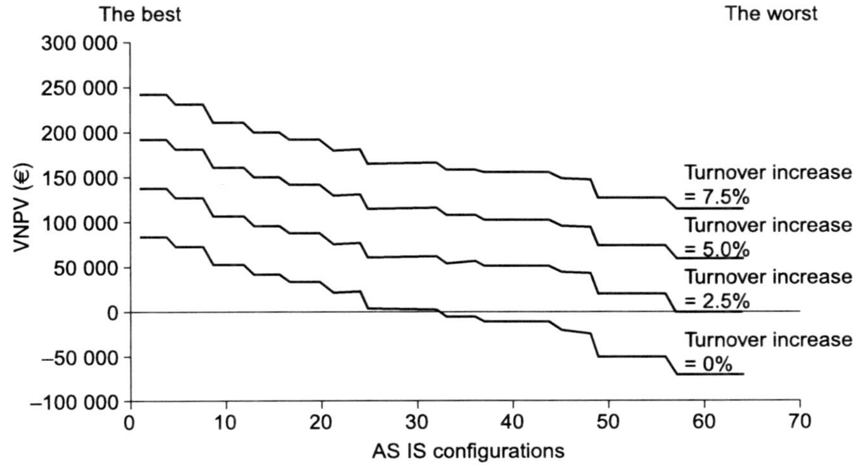

2.1 NPV over 5 years for a single retail store.

2.3.2 NPV of the investment for a representative supply chain

The scenario for a representative supply chain is composed of a DC and several RSs, whose number could range from 5 to 50. For the DC, all possible AS IS configurations are considered while for RSs, the best configuration is assumed. As before, the turnover increase is set at 0, 2.5, 5 and 7.5%. Figure 2.2 shows the outcomes for a turnover increase of 0%.

The analysis shows that economic outcomes are very different, depending on the supply chain structure examined and on the hypothesized value of turnover increase. If there is no increase in turnover (Fig. 2.2), the RFID investment generates a positive NPV when the number of RSs exceeds 25, irrespective of the AS IS configuration. This is an interesting result since, looking at Table 2.1, it can be appreciated that for 3 out of the 7 companies examined, the supply chain structure encompasses more than 25 RSs.

The achievable NPV varies substantially as a function of the number of RSs in the supply chain. A supply chain composed of 50 RSs benefits from a substantial NPV, ranging from about €3 500 000 (under the best configuration) to about €2 700 000 (under the worst configuration). Conversely, when considering fewer than 15 RSs, which is the case for one of the companies considered (Company 5), none of the AS IS configurations shows a positive NPV over the 5-year period; this result should be ascribed to the high cost of tagging incurred by the DC. For intermediate numbers of RSs, the profitability of the RFID investment depends on the specific AS IS configuration.

The number of RSs in the supply chain seems to have a significant impact on the NPV of the RFID investment; more precisely the profitability of the RFID

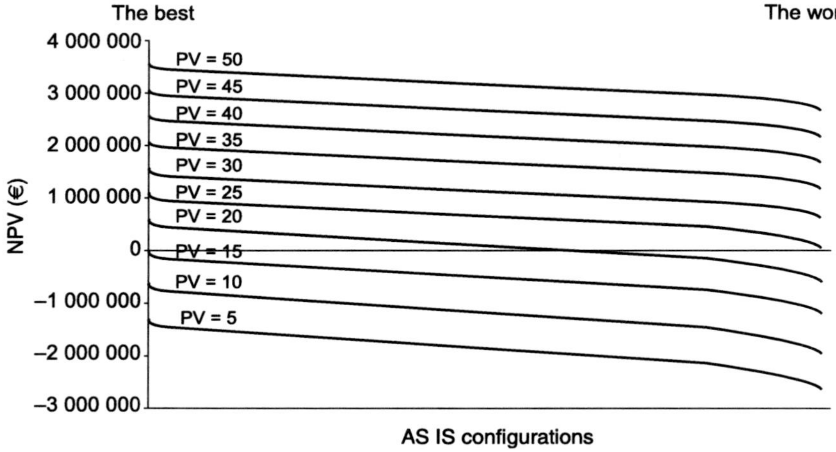

2.2 NPV of the investment for the supply chain (T.I = 0.0%).

investment becomes higher when considering increases in turnover of 2.5, 5 and 7.5%. When a 2.5% turnover increase is hypothesized, the RFID investment turns out to be always profitable if the supply chain is composed of more than 20 RSs; for turnover increases of 5 and 7.5%, it becomes profitable with more than 15 and 10 RSs, respectively. Conversely, if there are fewer than five RSs, the investment is always unprofitable.

2.3.3 Breakeven curve (BEC) of the RFID investment

The breakeven curve has been computed as a function of unitary tag cost (€/tag) and number of RSs in the supply chain; it is derived by identifying the combination of tag cost and number of RSs which makes the NPV of the investment nil over the 5-year period. The turnover increase is set at 0, 2.5, 5 and 7.5% as before; as an example, Fig. 2.3 shows the outcomes for a turnover increase of 0%. To be consistent with the previous results, the number of RSs considered ranges from 5 to 50 and the corresponding tag cost ranges from approximately 0.01 to 0.14 €/tag.

It is evident that investing in RFID is not always profitable when a turnover increase of 0% is hypothesized. Breakeven is reached under the best configuration for different values of tag cost and number of RSs. The boundaries of the BEC range from approximately 0.05 €/tag coupled with 5 RSs and 0.12 €/tag coupled with about 25 RSs. Depending on the AS IS scenario (corresponding to the internal portion of the graph), other combinations of number of RSs and tag cost allow the RFID investment to reach breakeven. Conversely, under the worst AS IS scenario, no combinations of number of RSs and tag cost allow the RFID investment to be paid back.

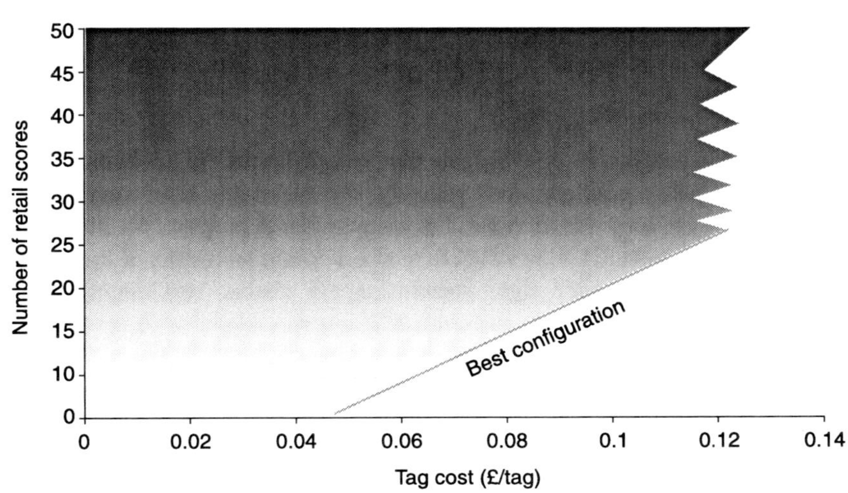

2.3 Breakeven curve of the RFID investment (T.I = 0.0%).

Results obtained when a turnover increase of 2.5% is hypothesized lead to similar conclusions. The RFID investment is again profitable under the best AS IS scenario, and the boundaries of the BEC approximate to a tag cost of 0.05 €/tag coupled with 5 RSs and 0.12 €/tag coupled with about 18 RSs. As per the previous case, under the worst configuration the RFID investment does not reach profitability for any combination of tag cost and number of RSs. Outcomes for turnover increases of 5 and 7.5% are slightly different, since the RFID investment can be reached for all AS IS configurations with a proper combination of tag cost and number of RSs.

When considering a 5% turnover increase under the best configuration, breakeven can be reached when the number of RSs ranges from 5 to 12, combined with a unitary tag cost ranging from about 0.06 €/tag to about 0.12 €/tag, respectively. Under the worst configuration, the boundaries of the BEC range from 15 to 45 RSs combined with a unitary tag cost ranging from about 0.02 €/tag to about 0.12 €/tag. For a 7.5% turnover increase under the best AS IS configuration, the boundaries of the BEC are approximately 5 to 12 RSs, combined with a unitary tag cost ranging from about 0.07 €/tag to about 0.12 €/tag, respectively. Under the worst configuration, the RFID investment reaches profitability with a minimum number of about 8 RSs, coupled with a tag cost of 0.03 €/tag.

Depending on the supply chain structure, the above results also make it possible to identify the tag cost that allows the RFID investment to reach breakeven. For instance, for a supply chain composed of 25 RSs and able to obtain a 5% increase in turnover from adopting RFID, the investment is profitable if the tag cost is lower than about 0.06 €/tag, while it is not profitable if the tag cost exceeds 0.13 €/tag. For intermediate values, the profitability of the RFID investment depends on the particular AS IS configuration of the supply chain.

2.3.4 NPV, IRR, PBP and ROI of some specific case studies

The aim of this analysis is to investigate the profitability of the RFID investment under the boundary configurations (namely, the 'best' and the 'worst'), with different values of turnover increase. For computational purpose, the number of RSs is set at 20. Table 2.4 shows the relevant economic indicators of investment profitability in the case of eight specific supply chains, resulting from the combination of different values of turnover increase, coupled with the best and worst configurations of DC and RS.

When considering the worst configurations of DC and RS, the RFID investment is highly unprofitable where the hypothetical increase in turnover is 5% or lower. In this situation, the supply chain experiences a relevant economic loss, ranging from about €4.3 to 1.2 million, at the end of the 5-year period. When a higher increase in turnover is hypothesized (i.e. 7.5%), the RFID investment reaches profitability at the end of the 5-year period. In particular, the NPV of the investment accounts for about €200 000, and the investment is paid back in 5 years.

When considering the best configurations of DC and RS, the RFID investment is always profitable over the 5-year period, and the NPV of the investment increases linearly with the percentage of turnover increase. The above results also suggest that, where the RS can benefit from a higher turnover increase (i.e. 7.5%), the RFID investment is always profitable, irrespective of the AS IS scenarios of DC and RS. In turn, this indicates that the turnover increase has a significant impact on the profitability of the investment, and should be taken into careful consideration in real cases.

2.4 Using RFID technology in practice: the RFID Fashion Pilot (RFP) project

The RFID Fashion Pilot (RFP) project was the final part of a research path started at the University of Parma RFID Lab in 2008. A group of ten companies (Benetton, Branded Apparel, Dolce & Gabbana Industria S.p.A. in partnership with DHL and TNT, Imax, Diffusione Tessile, Miroglio Fashion, Trussardi and ND Logistics) teamed up with RFID Lab academics in 2009 to start the RFP project. Although only one company provided the locations for in-field measurements and the garments (two sites of the Miroglio Fashion company), all participants shared both benefits and costs, were involved in project engineering, both from the technical and operational points of view, as well as in the experimental campaigns, and shared the resulting know-how.

A board of RFID Lab technology partners supplied state-of-the-art RFID hardware and software for the RFP project. The whole installation was fully deployed in January 2010, just in time for the 2010 spring/summer campaign. The flow of goods for spring/summer and fall/winter 2010 collections was monitored. The following garments were studied during the pilot study: jackets and coats (827 models); shirts, blouses and T-shirts (336); accessories (50); footwear and leather goods (43). Overall, 21 663 garments were tracked in the pilot supply chain, with the sites including a RS, located in a shopping district near Parma (Italy), and the corresponding supplier, a distribution centre (DC) located near Cuneo (Italy), at the worldwide logistics base of Miroglio Fashion group. Of the items monitored, 18 963 were sold. The in-field data collection for the spring/summer and fall/winter Miroglio collections started on 2 April 2010 and ended on 1 February 2011.

2.4.1 Tracking points and data collection

During the RFP project, the following processes were considered for RFID implementation:

- *DC processes*: packing and marking, palletizing of items, shipping; and
- *RS processes*: product receiving, replenishment from backroom, inventory counts, check-out.

The above processes were examined before RFID implementation (AS IS scenario) and after RFID reengineering (TO BE scenario), to assess the impact of introducing RFID technology. TO BE scenarios encompassing different RFID read points were engineered and deployed. The read points were termed 'tracking points', reflecting their ability to trace punctually the 'who', 'where', 'when', 'why' and 'how' of every item moving downstream in the supply chain. The whole deployment was based on EPCglobal standards, as well as on RFID item-level tagging, with the tag inserted in the product label. Items were tagged using EPC SGTIN product codes, and information records were captured in the EPCIS according to EPCglobal standards.

The first tracking point was located at the Miroglio Fashion DC, where individual clothing items and accessories were tagged during the packing and marking process. The price tag was printed using a converted tag, embedding an EPC Gen2 passive ultra-high frequency (UHF) RFID inlay. The price tags were applied as normal, the process being fully transparent for the operator. This process thus creates a unique association between the EPC serial and the product type, model, size, colour and price. If several items were aggregated in a case, the same association was created between the tag of the items aggregated and the tag of the case.

The second RFID tracking point was connected to the shipping process. The warehouse operator used either an RFID mobile device or a portal to cross-check every movable bin prepared for the RS in the pilot store against the list of expected items before loading it onto the truck, enabling the cases shipped by the DC to be certified.

The third RFID tracking occurred after shipping, when the products arrived at the RS, where the operator, on receiving the goods, input the bill of loading into a client terminal, and items or movable bins received at the store were thus automatically checked to make sure they corresponded with the expected delivery. This process could be performed by both the RFID handheld device and the replenishment door portal. In order to correctly input inventories, the operator needed to declare where the receiving process took place, either in the backroom or in the store. Thanks to the RFID reads, the store associate was able to punctually cross-check quantities for style/colour/size.

The fourth RFID tracking point was related to store replenishment, which consisted of restocking the items on the store shelves, and picking them from the backroom. A portal was installed in the aisle connecting the store floor with the backroom, and was equipped with four near-field antennae, in order to have a very selective reading field and avoid ghost reads. Items moved from the backroom to the store shelves were tracked when passing the RFID gate, and the corresponding inventory was moved from the backroom to the store, or vice versa, depending on the last known location of items in the store. Hence, the inventory count could provide information regarding the items available in both areas.

In the store area, four fitting rooms were equipped with RFID readers and near-field antennae, in order to be able to track what items were tried on by the customers and when. A second read point in the store area was deployed at the check-out counter. Here a fixed RFID reader read the RFID tags of items placed on the counter just before checking out. The same reader also managed a final read point deployed in a trash can, a box where the retained RFID tags of sold items were read.

RFID inventory counts were conducted regularly either in the store or in the backroom, and represented the last tracking point. A fixed RFID reader was adapted to work as a handheld device, but with a higher processing power to better handle reading large numbers of tags. Some processes typically occurred only once (e.g. tag encoding, shipping, receiving), while others could occur several times for the same item (e.g. inventory, fitting or replenishment).

2.4.2 Project results

The impact of RFID technology on the apparel supply chain can be manifold. In this section we present the many benefits that emerged from the RFP project, detailing the main numerical results obtained for every point.

Labour efficiency

An immediate benefit of RFID is the increased automation of internal processes, by enabling multiple, out-of-sight and completely automated identification of items. In the RFP project, we measured quantitatively the impact of RFID on labour productivity during the shipping process at the DC and receiving at the RS. The same savings were also assessed with respect to inventory counts.

For the shipping process at the Miroglio DC, we assessed the improvements in manpower productivity by comparing the time required for checking product cases during shipping before and after RFID implementation. In the AS IS scenario, employees used barcodes for product identification, and productivity during checks was approximately 7 to 8 barcodes checked per minute (i.e. 7–8 items/min). We found that, after RFID implementation, the time for checking decreased by 92 to 95%, resulting in a significant increase in manpower efficiency. For the receiving process in the RS, we used either the fixed RFID portal, which was also used for tracking replenishment from the backroom, or the RFID handheld devices. During receiving, we systematically measured the time required to read each movable bin using the mobile reader or the replenishment portal, and compared this value with the time needed to scan the items' barcodes. Finally, we converted all values into productivity expressed in items/min. During the pilot study, detailed data on the reading performance of RFID were collected for both simple and hanging garments, as a function of the RFID device used for the receiving process (handheld device or receiving portal).

Manual barcode scanning recorded an average productivity of 7.35 items/min. The main factor limiting barcode productivity is that the barcode may not be visible and has to be looked for. In the TO BE scenario, a movable bin could be read without removing hangers from the z-rack 97% of times using the RFID handheld device and 95% of times using the RFID portal. With respect to the overall reading accuracy, when hanging items were received through the RFID gate, RFID correctly received 84% of the movable bins, reaching a productivity of 52 items/min. Productivity increased to 62 items/min when the handheld device was used. In the case of folded items, manual barcode productivity did not significantly differ from hanging garments. With respect to the folded items received through the RFID gate, RFID productivity reached 77.64 items/min, and RFID provided 92% complete reads of the items included in the movable bin.

Sell-in accuracy

Sell-in accuracy is paramount in order to keep accurate inventories. Wrong sell-in input can lead either to under-estimated inventories (i.e. information system inventory is lower than the real one) or, even worse, to over-estimated inventories. While the former may lead to unnecessary reorders, involving excessive holding costs, the latter could potentially lead to stock-outs, and is therefore far more critical (Hardgrave *et al.*, 2009). During the RFP project, sell-in accuracy was evaluated through the post-processing of RFID data collected at store receiving and, in particular, by checking the correctness of the bill of loading (BoL). As already mentioned, in the AS IS receiving process, the store associates simply count the overall number of items received for every movable bin, and compare the amount with the corresponding numbers declared in the BoL.

Thanks to the deployment of RFID technology, in the TO BE process we were able to collect timely data on inbound flows received at the RS, and to compare items/quantities mix with the BoL. For each order line of BoL analyzed, we computed the resulting accuracy as the difference between the quantities in the BoL and those obtained from the RFID reads; 8 out of 46 transport documents (17%) contained errors in product quantity. The same analysis, carried out on order lines, showed that approximately 8.6% of them contained errors. It should be mentioned that, before the deployment of RFID, this issue was completely neglected by the store management, and the practice of relying on BoL data was widely accepted as not decreasing inbound productivity or inventory accuracy. Moreover, it is also important to underline that, without RFID deployment, it is difficult to detect errors in the BoL.

Inventory counts

Inventory counts are one of the principal uses driving the adoption of RFID technology in fashion and apparel. Thanks to RFID, it is possible to update physical inventories in the backroom or on the store floor and align information systems with real inventories, thus preventing stock-outs and reducing unnecessary holding costs due to inefficient replenishment.

During the project, we compared two distinct operative conditions for inventory counts:

1. A 'slow' (accurate) inventory count, which required 2 employees working for 30 min on average.
2. A 'quick' inventory count, which required an employee working for 7 min on average.

For the first option, we prioritized accuracy over productivity. We enrolled two operators: the first one holding the RFID handheld device and waving it over the items, the second one moving folded and lying garments in front of the handheld reader. This approach should have helped reduce missed reads due to tags being shielded or hidden. In the second option, we prioritized efficiency over accuracy and asked a single operator to move as fast as he could with a handheld reader and to collect as many reads as possible. We gave the operator approximately 7 minutes to complete this task.

Obviously, compared to quick inventory counts, the accurate ones are expected to reach a higher level of precision. The number of garments to be read was 4335 items. Slow counts gave an average of 4280 reads, corresponding to 98.73% accuracy and a productivity of 142 items/min. Conversely, the 3 quick inventory counts gave an average of 4168 items, corresponding to 96.16% and a productivity of 595 item/min each. By joining the 3 quick counts, we increased the accuracy to 97.83%. Finally, by joining the results of all inventory counts (i.e. 3 fast counts and the slow one), we reached a peak of 99.35% accuracy. From the figures, it is evident that the best practice for balancing accuracy and productivity is thus to encourage multiple quick inventory counts and combine the resulting RFID reads, rather than always trying to conduct accurate and slow inventory counts.

It should be stressed that, during an inventory count, it is in fact not possible to read all the 'expected' tags. This occurs because, for instance, items that have been stolen, while expected, cannot be read because they are not there. By cross-checking inventory data and RFID reads obtained at other tracking points, it was possible to discriminate between missed reads of items that were read somewhere else later, and missed reads of items that could not be read anymore. These latter items could have been either stolen or simply not detected by the system; whatever the reason, RFID allowed identifying some missing items, and highlighted the need for further checks. It is appropriate in this instance for an employee to verify whether the items were stolen or not detected; this could easily be done by comparing the RFID reads of this inventory count with subsequent ones.

Fitting

The RFID reader installed in the fitting rooms was equipped with a near-field antennae and put in continuous-read mode. This deployment made it possible to determine which items were selected by the customers, where those items were tried on, and how many try-ons they underwent. Before RFID deployment, such data were obviously unavailable. The analysis of the data collected indicated that 116 styles that have never been tried on were sold during the season. These styles were mainly accessories and could be considered as 'self-selling' items, since they did not require assistance from sales associates nor proper try-ons. Another 328 styles showed a ratio between number of sales and number of try-ons higher than 100%, meaning that more were sold than were tried on. A further 708 styles had ratios between number of sales and number of try-ons ranging from 100 to 0%. Probably such items did not fully meet the customers' expectations. The same happened for another 60 styles, which were taken to the fitting room but remained unsold.

Turnover increase

Assessing turnover increase was a primary objective for the RFP project. Sales were compared with a control store (i.e. a store that was not deploying RFID technology) to ensure that the increase in sales volume was not determined by differences such as store policies, weather conditions or location.

Participants in the RFP project suggested that RFID helped boost sales volume by reducing the number of times customers entering the store failed to find the item they wanted (because of the lack of the right style, size or colour). A specific RFID application was designed to provide sales assistants with the information they required to answer a specific customer request. If a customer asked for a particular style in a different size or colour, the sales assistant could use a fixed RFID reader to get information on a touch screen about different sizes and colours of the item, and whether they were available in the store or in the backroom. Similarly, the sales assistant could load the unique serial number of the sample item in a RFID-enabled handheld computer, to quickly locate it in the backroom.

To assess the impact of this application on sales volume, its use was tracked during the RFP project, and cross-checked with the check-out data. The application was used 996 times, and about 144 of the items that were looked for using the application were subsequently sold. Hypothesizing that, without RFID deployment, the sales associate would be unable to find the item requested by the customer (and therefore, the sale would be lost), the use of the application generated a 0.8% increase in total sales for the store. However, not all items found can be considered as sold as a result of being found, meaning that the use of the RFID application does not directly imply that the item has been sold. Hence a correction factor was derived by correlating the moment where the sale occurs with the use of the RFID application. For instance, hypothesizing that only items sold within an hour of the interaction can be linked to the RFID application, the correction factor scores 0.66 and the estimated sales increase is $0.8 \times 0.66 = 0.52\%$.

A second application was developed to optimize replenishment, and in particular to help store associates maintain an adequate assortment of all top-selling styles on the sales floor to counteract the problem of lost sales when popular items are not restocked quickly enough or forgotten in the backroom. Such items may end up being sold during the sales season at bargain prices instead of in season at full price. The dashboard developed showed the employees what styles, colours and sizes needed to be replenished on the store floor in real time; the application also highlighted items that were stocked in the backroom but were unavailable on the store floor. The system then tracked when the list was printed and the time when listed items were moved from the backroom to the store floor.

During the RFP, the replenishment application was used 30 times (i.e. on 30 days). On those days, sales totalled 1994 items, store associates replenished 239 items, and 221 items out of those replenished were sold. This generated an impact on sales volume of approximately 11.1%. The replenishment time should be correlated with the check-out time to identify correctly the impact of this RFID application on the sales increase. The impact of RFID on store turnover can be derived as a function of the dwell time: assuming that only items with a dwell time of a week (or less) contribute to sales volume, the increase in turnover due to RFID would be 4.91% of sales; if the dwell time is lower than 24 hours, the sales increase would be around 1%, which is still relevant to the fashion industry.

2.5 Sources of further information and advice

A complementary topic, related to efficient RFID deployment in the fashion supply chain, is the design of RFID tags specifically for fashion items. This topic focuses on the technical characteristics required of RFID technology for its successful application in the fashion context, and has been addressed by Bertolini *et al.* (2012). The analysis described in this chapter does not take into account the use of RFID as an anti-theft, or EAS, device. Bottani *et al.* (2012) explored the potential of implementing UHF Class 1 Gen 2 RFID technology to support EAS applications in the apparel supply chain. The authors analyzed the technical performance achievable using RFID technology under different theft patterns, and also the economic impact of adopting RFID for anti-theft purposes.

2.6 Acknowledgements

This chapter has been built by reprinting parts of the following publications:

- *International Journal of RF Technologies: Research and Applications*, 1(4), The impact of RFID technology and EPC system on logistics processes of the fashion industry supply chain, by Bottani, E., Ferretti, G., Montanari, R. and Rizzi, A., 225–52, © 2009, with permission from IOS Press.

- *International Journal of RF Technologies: Research and Applications*, 3(4), Experimental evaluation of business impacts of RFID in apparel and retail supply chain, by Bertolini, M., Bottani, E., Ferretti, G., Rizzi, A. and Volpi, A., 257–82, © 2011/2012, with permission from IOS Press.

2.7 References

Bertolini, M., Bottani, E., Ferretti, G., Montanari, R. and Volpi, A. (2012), Analysis of the requirements of RFID tags for efficient fashion supply chain management, *International Journal of RF Technologies: Research and Applications*, 3(1), 39–65.

Böse, I. and Pal, R. (2005), Auto-ID: Managing anything, anywhere, anytime in the supply chain, *Communications of the ACM*, 48(8), 101–6.

Bottani, E., Ferretti, G., Montanari, R., Rizzi, A. and Volpi, A. (2012), Performances of RFID, acousto-magnetic and radio frequency technologies for Electronic Article Surveillance in the apparel industry in Europe: A quantitative study, *International Journal of RF Technologies: Research and Applications*, 3(2), 137–58.

Bruce, M., Daly, L. and Towers, N. (2004), Lean or agile. A solution for Supply Chain Management in the textile and clothing industry, *International Journal of Operations and Production Management*, 24(2), 151–70.

Brun, A., Caniato, F., Caridi, M., Castelli, C., Miragliotta, G. *et al.* (2008), Logistics and supply chain management in luxury fashion retail: Empirical investigation of Italian firms, *International Journal of Production Economics*, 114, 554–70.

Choi, T. (2007), Pre-season stocking and pricing decisions for fashion retailers with multiple information updating, *International Journal of Production Economics*, 106(1), 146–70.

Choy, K. L., Chow, K. H., Moon, K. L., Zeng, X., Lau, H. C. W. *et al.* (2009), A RFID-case-based sample management system for fashion product development. *Engineering Applications of Artificial Intelligence*, 22(6), 882–96.

Christopher, M., Lowson, R. H. and Peck, H. (2004), Creating agile supply chains in the fashion industry, *International Journal of Retail and Distribution Management*, 32(8), 367–76.

Christopher, M. and Towill, D. (2001), An integrated model for the design of agile supply chains, *International Journal of Physical Distribution & Logistics Management*, 31(4), 235–46.

EPCglobal (2004), The EPCglobal Network™: Overview of design, benefits and security. Available from: http://www.epcglobalinc.org (accessed July 2006).

EPCglobal (2007), EPC Information Services (EPCIS) Version 1.0 Specification, Ratified Standard. Available from: http://www.epcglobalinc.org (accessed April 2008).

Evans, J. R. and Berman, B. (2001), Conceptualizing and operationalizing the business-to-business value chain, *Industrial Marketing Management*, 30(2), 135–48.

Hammer, M. and Champy, C. (1993), *Reengineering the Corporation*, New York: Harper Business.

Hardgrave, B. C., Aloysius, J., Goyal, S. and Spencer, J. G. (2009), Does RFID improve inventory accuracy? A preliminary analysis, *International Journal of RF Technologies: Research and Applications*, 1(1), 44–56.

Harrison, A., Christopher, M. and van Hoek, R. (1999), *Creating the Agile Supply Chain*, London: Institute of Logistics & Transport.

Hines, T. and McGowan, P. (2005), Supply chain strategies in the UK fashion industry – The rhetoric of partnership and realities of power, *The International Entrepreneurship and Management Journal*, 1(4), 519–37.

Jacobs, D. (2006), The promise of demand chain management in fashion, *Journal of Fashion Marketing and Management*, 10(1), 84–96.

Lo, W.-S., Hong, T.-P. and Jeng, R. (2008), A framework of E-SCM multi-agent systems in the fashion industry, *International Journal of Production Economics*, 114, 594–614.

Loebbecke, C. (2005), RFID technology and applications in the retail supply chain: The early Metro Group pilot, *18th Bled Conference Integration in Action*, 1–11.

Lowson, R. H., King, R. and Hunter, N. A. (1999), *Quick Response: Managing the Supply Chain to Meet Consumer Demand*, Chichester: Wiley.

Moon, K. L. and Ngai, E. W. T. (2008), The adoption of RFID in fashion retailing: A business value-added framework, *Industrial Management & Data Systems*, 108(5), 596–612.

Nair, A. and Closs, D. J. (2006), An examination of the impact of coordinating supply chain policies and price markdowns on short life-cycle product retail performance, *International Journal of Production Economics*, 102(2), 379–92.

Reynolds, K. E., Folse, J. A. G. and Jones, M. A. (2006), Search regret: Antecedents and consequences, *Journal of Retailing*, 82(4), 339–48.

RFID Journal, 12 December 2008, American Apparel expands RFID to additional stores. Available from: http://www.rfidjournal.com/article/articleview/4510 (accessed January 2009).

RFID Journal, 14 October 2009, GS1 Releases Guidelines for RFID-based Electronic Article Surveillance. Available from http://www.rfidjournal.com/article/pdf/5303/1/1/rfidjournal-article5303.PDF.

Roy, S., Sivakumar, K. and Wilkinson, I. F. (2004), Innovation generation in supply chain relationships: A conceptual model and research propositions, *Journal of the Academy of Marketing Science*, 32(1), 61–79.

Taylor, R. R., Morgan, R. M. and Morton, A. R. (2003), Efficient versus responsive supply chain choice: An empirical examination of influential factors, *Journal of Product Innovation Management*, 20(6), 430–43.

Zahay, D. L. and Handfield, R. B. (2004), The role of learning and technical capabilities in predicting adoption of B2B technologies, *Industrial Marketing Management*, 33(7), 627–41.

3

The role of radio frequency identification (RFID) technologies in improving process management and product tracking in the textiles and fashion supply chain

S. G. AZEVEDO, University of Beira Interior, Portugal and P. PRATA and P. FAZENDEIRO, Instituto de Telecomunicações, Portugal

DOI: 10.1533/9780857098115.42

Abstract: This chapter aims to gain a better understanding of RFID deployment in the Fashion and Textile Supply Chain (FTSC), mainly concerning its application in supporting operations, its advantages and main business drivers. In an attempt to find an RFID deployment pattern within the FTSC, a cross-case analysis was performed. The RFID experience of five companies based in different countries was analysed. From these empirical data, three main conclusions are drawn: (1) RFID is deployed mainly to support the following operations: the handling process; tracking work-in-progress; receiving operations; shipping operations; tracking products; tracking inventories; monitoring and sorting of merchandise; counting stock and picking merchandise; tracking containers; shipping; locating products; and store management. (2) The main advantages highlighted are: better inventory management; improved read rates; and order accuracy. (3) The main driver that led the case study companies to deploy RFID technology was the identification of inefficiencies in several of their operations.

Key words: RFID, fashion and textile supply chain (FTSC), industry operations, business drivers, case study.

3.1 Introduction

Radio Frequency Identification (RFID) technology is a wireless identification technology that improves communication of electronic information associated with physical items (Modrak *et al.*, 2010). It is composed of several elements (Atkinson, 2004):

- readers
- tags
- software
- security programs for readers.

RFID technology is an automatic identification system that uses radio frequency signals for wireless communication (Khan *et al.*, 2009) and was patented in the

1980s by Charles Walton (1983). Its potential for the reduction of inventory losses, increases in efficiency and speed of processes, and the improvement of information accuracy (Sarac *et al.*, 2010), makes it an important tool for organizations and supply chains in sustaining their competitive advantage (Erickson and Kelly, 2007; Mehrjerdi, 2010; Visich *et al.*, 2009). Indeed, RFID has become an important source of competitive advantage by increasing supply chain transparency through the dissemination of large amounts of accurate and real-time data to enhance decision-making processes.

The value of RFID technology is particularly visible in Fashion and Textile Supply Chains (FTSCs), as the fashion business is characterized by a wide assortment of products, short life-cycles, high seasonality, high volatility, high-impulse purchasing and complicated distribution and logistics operations (Castelli and Brun, 2010; Christopher *et al.*, 2004). This FTSC business model puts significant emphasis on rapid delivery, high-quality products and low prices for each supply chain segment. In this highly competitive environment, the availability of a certain product category (or a specific item) is a relevant source of value to customers, and hence there are huge penalties associated with the current season's 'must-have' advertised items being out of stock (Brun and Castelli, 2008). To respond to customer needs, 'time compression' and flexibility should be developed into the whole supply chain, beginning with the textile product itself. Castelli and Brun (2010) also stress the need to align operations of different FTSC members along the supply chain, enhancing information exchange through the use of new communication tools and process coordination practices. As stated by Lo *et al.* (2008), building an integrated supply chain management information system to combine information and share knowledge of the entire supply chain is an important task to improve the efficiency and effectiveness of its management.

This technology is now at a stage where there are potentially larger advantages from a wider range of applications but barriers still remain. Despite numerous studies on the advantages of RFID technology for several supply chain operations, Lee and Ozer (2007) observe that a large gap exists between the perception and reality of what is generally assumed to be the benefits engendered by RFID deployment. According to Kapoor *et al.* (2009), there is a lack of in-depth studies considering the drawbacks of RFID implementation in a supply chain management context. In addition, Sarac *et al.* (2010) stress the need to conduct research for multiple items and multiple actors in order to provide more realistic analyses of supply chains.

In this chapter, a cross-case analysis is performed applying investigator triangulation (Jick, 1979) to gain a better understanding of RFID deployment in the FTSC. The focus is mainly on its application in supporting companies' operations, and on the advantages and drivers for adoption of the technology. The remainder of this chapter is organized as follows. The general application context of RFID technology in a supply chain is presented. This is followed by a discussion of its enhanced application to FTSC operations, along with the derivation of a set of working propositions. In order to assert the validity of such propositions, a cross-case analysis is then presented, with a description of the methodology employed and a presentation of the case studies analysed. The main conclusions of the study are drawn in the final section of the chapter.

3.2 Using RFID in supply chains: an overview

The widespread use of RFID in supply chain management only started in the last decade, when a number of world leading retailers (including the Metro Group in Germany, Marks & Spencer in the UK and Wal-Mart in the USA) mandated the use of RFID for tracking supplies (*Cover Pages*, 2003; Roberti, 2003). According to a study from the RFID Research Center at the University of Arkansas (Cromhout *et al.*, 2012), in the first stage of RFID usage (2003–2006), the focus was to identify pallets and cases moving from distribution centres to retail stores. In the second stage (2006–2009), the focus shifted to individual items in store. Finally, in 2010, the main conclusion was that RFID should be used in the entire supply chain from point of manufacture to store shelf.

In 2010, the Voluntary Interindustry Commerce Solutions (VICS) association launched the Item Level RFID Initiative, with the purpose of establishing a course of action for electronic product code (EPC) adoption throughout the retail supply chain (ILRI, 2010). EPC-enabled RFID technology was designed to allow the unique identification of every single instance of each item manufactured, as opposed to just the manufacturer and class of products, as barcodes usually do.

The moment to incorporate the concepts of full supply-chain visibility into business philosophy is when RFID technology is already stable and its widespread use has led to a drastic reduction in the price of hardware and software (Javick, 2011). In Ustundag (2013), RFID is presented as a key strategic technology that adds value throughout the supply chain by increasing process efficiency, accuracy, visibility and security. This means that RFID investment should be seen as a strategic business decision and not as a technological choice.

In the last few years, there have been high levels of investment in the development and improvement of RFID systems, because of the important advantages that companies and supply chains can achieve in their operations management, when compared to barcodes (Chang *et al.*, 2010). These advantages derive mainly from innovations in identification operations, and then from the reduction or even elimination of the manual work required to scan items.

This system also has a high capacity, whilst being simultaneously rigorous (So and Liu, 2006), which could become an important source of competitive advantage and innovation in the operations management field. These characteristics contribute to increasing the efficiency of a range of operations such as materials displacement, for example, since warehouse operators do not have to manipulate an optical reader to collect data about the products that are being loaded or unloaded. Moreover, this new technology allows the collection of information from objects on the move (Knill, 2002), enhancing the visibility of material flows and allowing better coordination between supply chain entities (Chang *et al.*, 2010).

RFID also allows the tracking of every single article along the FTSC. A unique EPC identifier can be written into the tag applied to each article, and the specific product features, along with supply chain information, can be associated with each identifier in a firm's information system (Mogre *et al.*, 2009). To Gaukler and Seifert (2007), the main advantages resulting from RFID deployment in supply chains operations management can be categorized into labour and time savings and also in the benefits from increased visibility. According to Mogre *et al.* (2009), this visibility enables new replenishment policies such as lateral trans-shipment, as with the shipment of goods between two retail stores in order to reduce stock outages, inventories and transportation costs. Sarac *et al.* (2010) analysed the advantages of RFID in overcoming inventory inaccuracy, the bullwhip effect and replenishment policies; these and other advantages are identified in Table 3.1.

Table 3.1 Main advantages of RFID across the supply chain

Manufacturers	**Inventory** • More information on inventory (Jones *et al.*, 2004) • To decrease stock levels (Kelly and Scott, 2005) • To decrease stock outs (Kinsella and Elliot, 2005; Veeramani *et al.*, 2008) **Operational** • Production and supply availability (Karkkainen and Holmstrom, 2002) • Better quality control during production (Zhou and Piramuthu, 2012) • Better monitoring of work-in-progress and finished products (Moon and Ngai, 2008) **Information availability** • Quicker data collection (Veeramani *et al.* 2008; Witt, 2006) • Quicker identification of products (Borck, 2006)
Distributors/ logistics providers	**Inventory** • To decrease stock outs (Veeramani *et al.*, 2008) • A more accurate inventory, reduced shrinkage and theft, and more efficient use of storage space (Jones *et al.*, 2005) **Information availability** • Quicker data collection (Veeramani *et al.*, 2008; Witt, 2006) • To reduce the number of incorrect manual counts, mislabelling, and inaccessible/misplaced inventory (Kelly and Scott, 2005; Kinsella and Elliot, 2005) **Delivery** • To minimize distribution errors (Moon and Ngai, 2008) • To improve space use in warehouse and distribution centres' space (Albright, 2005) • To decrease charge back (Veeramani *et al.*, 2008) • Deliveries are authenticated at a higher level of accuracy and reduction in human error in counting (So and Liu, 2006)

(*Continued*)

Table 3.1 Continued

Retailers	**Service level** • Better promotion of products and stimulated up selling (Atkinson, 2004; Kelly and Scott, 2005; Moon and Ngai, 2008) • To trigger an interactive display of related products (Jones *et al.*, 2005) • Easier target of customers (Jones *et al.*, 2004; Moon and Ngai, 2008) • To improve customer service (Roberti, 2003) • To support customers making decisions on how to mix and match various clothes within a few minutes (Kwok and Wu, 2009). • To capture customer preferences and monitor inventory at each POS in real time (Kwok and Wu, 2009) **Inventory** • Better track of products that are selling rapidly (Jones *et al.*, 2004) • To improve inventory management (Moon and Ngai, 2008; So and Sun, 2010) • To reduce misplacement (Moon and Ngai, 2008; So and Sun, 2010) • To improve process visibility (So and Sun, 2010) • Enables the deployment of lateral trans-shipment (the shipment of goods between two retail stores) (Mogre *et al.*, 2009). **Information availability** • To reduce human errors (Moon and Ngai, 2008) • Quicker data collection (Moon and Ngai, 2008; Veeramani *et al.*, 2008; Witt, 2006) **Security** • To reduce shrinkage and theft (Coltman *et al.*, 2008; Grant and Fernie, 2008; So and Sun, 2010) • To reduce damage (So and Sun, 2010) • To improve security control (So and Sun, 2010)

3.3 The use of RFID technology in fashion and textile supply chains (FTSC)

Firms in the textile and fashion industry have been regarded as late adopters of innovative technologies, lagging far behind firms in other sectors such as the automotive, healthcare and electronics industries (Moon and Ngai, 2008). Short product life-cycle, long production lead time and high forecasting errors are critical problems found in this specific supply chain (Lam and Postle, 2006). Bruce *et al.* (2004) identified low profit margins for domestic suppliers, slow responsiveness to a competitive market place and high product variety as additional problems.

The FTSC involves raw material suppliers, textile manufacturers, clothing manufacturers, retailers and, finally, consumers. This kind of supply chain deploys

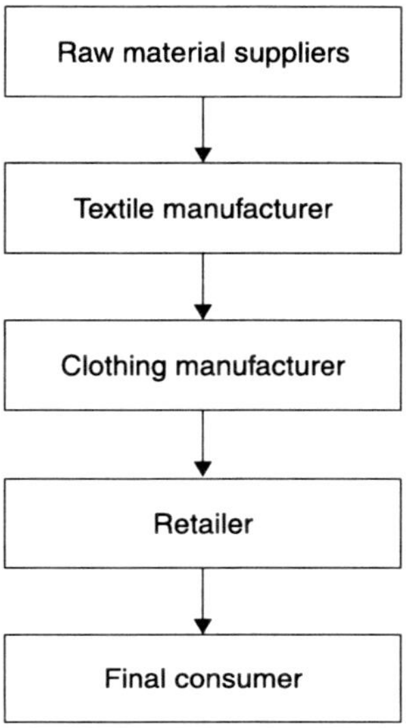

3.1 Fashion and textile supply chain.

a pull strategy, using all data on consumer behaviour as strategic information (Fig. 3.1). The Fashion and Textile industry is characterized by labour-intensive and multi-faceted processes with relative technological simplicity (Azuma *et al.*, 2009). The goals of supply chain management in this kind of industry are delivering the textile and fashion trends at the right time in the right place, with increased variety, affordability and customization, thus satisfying both existing and potential customer needs (Azuma *et al.*, 2009). To maintain competitiveness in the global market, manufacturers need to both shorten the lead time of new products and increase their responsiveness to market changes, using approaches and tools to automate their logistics operations and improve their access to information (Bottani, 2009).

Bruce *et al.* (2004) argue that keeping low product inventories, adopting technology to manage product replenishment and delivery processes, sourcing overseas and manufacturing domestically could all save costs, reduce lead times and provide quick responses to customer demands. Fashion products constantly change and induce companies in the FTSC to introduce internal changes; So and Sun (2010) suggest that applying the lean services used in apparel retail operations across the board may equally benefit FTSC companies. The use of RFID technology may have a significant impact on every link along the supply chain (Chen and Pfleuger, 2008; Melski *et al.*, 2007).

The deployment of RFID in the FTSC helps ensure more efficient supply chain processes (Gaukler and Seifert, 2007; Loebbecke, 2004), providing higher transparency throughout the supply chain from textile manufacturers to fashion retail stores (Chappell *et al.*, 2003b), as well as new options for advertising inside stores (Chappell *et al.*, 2003a). Hinkka *et al.* (2010) discuss the FTSC characteristics that make it appropriate for RFID utilization: product variety is high and the different products are difficult to separate at first sight. Another issue in FTSC is the different needs of each supply chain member. Fashion retailers face the challenge of adapting quickly and frequently to keep up with product changes and demand trends, to meet customer needs (Bottani, 2009). In this regard, the ability of the RFID to track and trace items can streamline many operations along the supply chain. However, Hinkka *et al.* (2010) argue that RFID solutions are mainly designed for retail purposes, even though the same tags could be used upstream in the supply chain.

Gaukler and Seifert (2007) argue that assembly and manufacturing operations could benefit from the advantages associated with RFID: tags could be used in a manufacturing setting to identify the product that is being produced, allowing savings in labour and reworking costs. If tags are attached early in the manufacturing phase, other supply chain members could also exploit them, increasing supply chain visibility and potentially reducing handling errors and confusion. RFID advantages could also be realized in shipment handling, as well as in picking and packing processes (Bensel *et al.*, 2008). Gaukler and Seifert (2007) also discuss the usefulness of RFID in logistics activities across a supply chain, namely in reducing bottlenecks, enabling a faster and less costly product movement, and improving inventory accuracy.

In retailing, improvements in on-shelf availability of goods can be achieved by redesigning the shelf replenishment process, creating transparency within the actual inventory in store and, in this way, improving inventory control (Bensel *et al.*, 2008). Moon and Ngai (2008) suggested that the use of RFID technology in retail can streamline stocktaking operations, strengthen customer relationships, facilitate promotional activities and allow retailers to allocate resources more effectively. Table 3.2 illustrates RFID deployment to support different operations across a supply chain.

This overview suggests the following proposition:

P_1: Companies in the FTSC can deploy RFID technology to support operations.

The increased deployment of RFID technology in the FTSC is motivated essentially by the following advantages that companies can expect to achieve: increasing sales, improved efficiency, better flexibility, improvements in forecasting and planning, error reduction, traceability, workflow improvement, lead time reduction, internal inventory control, production status, improved shelf management, enhancement of customer service, responsiveness, reduction in the time spent counting, cost savings, reduction in shrinkage, increased visibility, labour reduction, and a shorter order-to-cash cycle. Many of these advantages are related to inventory measures (Jones *et al.*, 2005; Kinsella and Elliot, 2005; Veeramani *et al.*, 2008).

Table 3.2 Operations supported by RFID across the FTSC

Manufacturer / supplier	• Packaging at production line end (Miragliotta *et al.*, 2009) • Receiving shipments (Miragliotta *et al.*, 2009) • Storage (Miragliotta *et al.*, 2009) • Inventory control and management (Miragliotta *et al.*, 2009) • Shipping (Miragliotta *et al.*, 2009) • Item tracking within manufacturing plant (Butcher, 2007)
Distributor/ logistics providers	• Packaging (Bottani, 2009) • Inventory control (Bottani 2009; Hinkka *et al.* 2010; Miragliotta *et al.*, 2009) • Shipping and item tracking (Baars *et al.*, 2009; Bottani, 2009; Miragliotta *et al.*, 2009) • Shipping consolidation loading (Butcher, 2007) • Picking (Baars *et al.*, 2009; Bottani, 2009) • Receiving shipments (Bottani, 2009; Miragliotta *et al.*, 2009) • Tracking of reusable packaging (Baars *et al.*, 2009) • Verification (Butcher, 2007) • Conveyance loading (Butcher, 2007) • Conveyance tracking (Butcher, 2007) • Order assembly (Miragliotta *et al.*, 2009)
Fashion retailers	• Receiving shipments (Bottani, 2009; Miragliotta *et al.*, 2009;) • Storage (Bottani, 2009; Butcher, 2007; Miragliotta *et al.*, 2009) • Inventory control (Hinkka *et al.*, 2010; Miragliotta *et al.*, 2009; Moon and Ngai, 2008) • Out-of-stock control (Miragliotta *et al.*, 2009; Mogre *et al.*, 2009) • Shop floor control (Moon and Ngai, 2008)

The RFID literature suggests a second proposition:

P_2: Most of the RFID advantages recognized by companies in the FTSC are related to inventory management.

In Roussos (2008), a set of business drivers for implementing RFID technology for supply chain optimization is identified:

- *Handling efficiency*: resulting mainly from the fact that RFID allows the automatic identification of product and containers without the need of a line-of-sight between the reader and tag, unlike with barcodes.
- *Out-of-stock reduction*: RFID can increase product availability by reducing the number of delivery errors and by increasing inventory accuracy in all stages of the supply chain.
- *Inventory reduction*: the use of RFID may increase the accuracy and timeliness of demand forecasts.
- *Order reconciliation*: this technology can prevent delivery errors, facilitate delivery confirmation and can be a means of proof for shipment delivery.
- *Theft prevention*: RFID at the item level works as an anti-theft system. RFID may also be useful in preventing the introduction of counterfeit trade throughout the supply chain.

Moreover, according to Charikleia (2010), the main forces that drive retailing companies to use RFID technology are:

- *The need for total visibility*: RFID can give visibility throughout the supply chain and thus minimize products or supplies shortages.
- *Increased information sharing*: RFID can increase collaboration between the different departments within a company, as well as between supply chain partners.
- *Increased coordination*: RFID can enable successful coordination between members of a supply chain, in terms of strategies and decision-making.
- *Increased competition*: companies need to keep up with new technologies and concepts to face market conditions and RFID is considered as a good solution to improve supply chain efficiency.
- *Increased customer awareness and demands*: today's customers have high demands and expectations and RFID can enable the full control of each individual product.
- *Technology trigger*: the wide advertising of RFID technology and extensive research in the area make it an attractive option for a large number of companies.
- *The continuously growing need for cost reduction*: RFID appears to be an attractive solution to optimize every business process, achieving lower costs and increasing efficiency throughout the supply chain.

There are some specific drivers promoting adhesion to RFID downstream of the FTSC, where inventory is constantly on the move – between shelves, the dressing room and the backroom – unlike in other retail segments where items have a set location on the rack. The study presented in Waller *et al.* (2011) identifies a set of potential use cases where drivers to the implementation of RFID can be identified:

- *Store operations*: improving backroom-to-shelf replacement and improving perpetual inventory accuracy are the processes that can have the most benefits for apparel manufactures (Delen *et al.*, 2007; Gaukler and Seifert, 2007; Hardgrave *et al.*, 2008).
- *Demand signals*: the main driver is to improve the accuracy of order forecasts (Williams and Waller, 2010, 2011).
- *Inventory management*: the main drivers for using RFID are related to the reduction of both cycle and safety stock.
- *Distribution centre (DC) operations*: RFID tags can be used to improve cycle counting in speed and accuracy and can also have a positive impact on pick/pack operations.
- *Inbound and outbound processes*: item-level RFID tagging offers the ability to audit the content of each carton through an automated scan process, reducing the time and labour in inbound and outbound processes.
- *Tracking and traceability shipping*: RFID tracking and tracing enable quality management and continuous quality improvement.

In recent years, the use of RFID technology has also been seen as suitable for identifying and preventing counterfeiting in the fashion supply chain. In Miragliotta *et al.* (2010) and Amato *et al.* (2012), models to evaluate how RFID technologies can help to prevent product counterfeiting and grey markets are presented.

As can be seen by this literature review, many of the drivers suggested for deployment of RFID technology by companies are associated with inefficiencies in many different operational processes. This suggests the following third proposition:

> P_3: The main driver for companies in the FTSC to deploy RFID technology is to overcome inefficiencies in operational processes.

3.4 Using RFID technology in practice: the use of case studies

The main objective of this research is to investigate RFID technology deployment in the FTSC in order to identify the main operations supported by it, and also look at the advantages and drivers associated with it. As there is sometimes a gap between management research and practice (Markides, 2007; Shapiro *et al.*, 2007), a case study approach was selected in an attempt to fill this gap. According to Ellram (1996), case studies are the best means to understanding a certain phenomenon, as they provide depth and richness, allowing the researcher to really understand the what, how and why questions pertaining to a given situation. From both Perry's (1998) and Rowley's (2002) points of view, a case-study approach is adequate when the boundaries of a phenomenon are not only unclear but there is also no control over behavioural events, as in this case. Attacking the same problem with a variety of methods from a variety of angles can be useful, not only for the study itself, but also for the validity of the analysis. If the same phenomenon is observed with more than one method it will, most probably, increase the authenticity of the observation.

Bearing in mind this advice, in this research, RFID deployment in FTSC is explored from the point of view of RFID users (textile manufacturers, clothing manufacturer, logistics providers and retailers). According to Oppermann (2000), using a team of diverse researchers may help to reduce any eventual bias and, with this in mind, this research team was composed of three researchers: two experts in computer science and engineering, and one in management and logistics. Furthermore, using multiple data sources may lead to discoveries that would not have been made otherwise. To this end a triangulation approach can be used, combining multiple approaches to 'zero in' on the answers to a research question (Singleton and Straits, 1999). This can involve combining multiple data sources (data triangulation), using multiple research methods to analyze the same problem (methodological triangulation), or using multiple investigators to work on the same task (investigator triangulation) (Jick, 1979; Oppermann, 2000). In this chapter, data and investigator triangulations were used, as explained below. Triangulation allows the researcher to address a broader range of issues, including historical, attitudinal and behavioural factors (Yin, 2002).

The case-study method has three distinct stages: design, collection and analysis. The final stage is an analysis of the individual case studies, allowing 'cross-case' reports to be written (Yin, 2002). In general, the search for patterns can be conducted using internal or cross-case analyses. When using more than one case in a study, it is logical to find similarities and differences across the cases using various cross-case analysis tools. Among others, this practice can help to identify critical predictor variables and causal interactions, and to validate or extend existing theoretical models (Stuart *et al.*, 1998). Thus, pattern-matching is rightly characterized as a form of empirical validation for qualitative data (Johnston *et al.*, 1999).

To reach this chapter's objectives, five illustrative case studies were conducted based on secondary data using external sources, namely books, journals, business magazines and websites. Despite being limited to the selected case studies and to the available data in external sources, this research design helps to define issues specifically before a primary study, such as an in-depth case study, is undertaken. The primary data will support the validity of the propositions. To this end, the case studies were analyzed as a means to:

- identify the main operations supported by RFID, the advantages and drivers most referred to by companies positioned at different FTSC levels; and
- to obtain a deeper knowledge of the way different partners deal with RFID, i.e. which operations, advantages and drivers are identified by them.

Using a methodology similar to Finch (2004), McKinnon (2006) and Visich *et al.* (2009), the secondary data for this research study were gathered from the analysis of published literature based on a broad range of sources including newspapers, conference proceedings, industry reports, White Papers, press releases and books. In addition, some specialized magazines on RFID, such as *RFID Journal* and *Logistics Today*, were used. To take part in this research, RFID companies had to:

- belong to an FTSC; and
- deploy RFID technology.

The selected articles describing case studies were analyzed and, finally, individual aspects are briefly described with a special focus on operations supported, and the advantages and drivers of RFID technology deployment in FTSC. The objective was not to offer further insight into the single case studies, but to bring them together to get a wider picture and learn from the cross-case analysis (Seuring, 2004).

3.5 Case studies in the use of RFID technology: background and results

A set of illustrative case studies, consisting of five international companies belonging to different nodes of the supply chain, were analyzed to illustrate RFID deployment in the FTSC. Each of the companies' experience with the RFID system was explored to identify the operations where RFID is most deployed, the main advantages reached with RFID implementation, and the main drivers that motivate its deployment. The objective was to give answers to the propositions formulated in the previous section and to reach a deeper knowledge on the deployment of RFID in the FTSC. In order to uncover contextual issues, exploratory case studies were developed (Yin, 2002). Data was gathered from newspapers, conference proceedings, industrial reports, White Papers, press releases and books and also some journals and magazines on the topic. This section is organized as follows: first, the sample profile is presented; second, individual case analyses are developed to identify the main operations supported by RFID, and the main advantages and drivers that motivate the case study companies to deploy RFID.

3.5.1 Summary of the case studies

The case studies used were: one textile manufacturer (Griva); one clothing manufacturer (Gardeur AG); one clothing and fashion retailer (Lemmi Fashion); one fashion retailer (Kaufhof); and one logistics service provider (DHL Solutions Fashion). Company profiles, locations and company sizes are presented in Table 3.3, whilst the operations in which RFID were deployed, the main drivers to RFID utilization and the main advantages reached with it, are described for each individual company in Table 3.4.

As can be seen in Table 3.3, the case studies are taken from several countries: Italy, Germany, and also France. This allows a wider comparison and a better understanding of RFID deployment in FTSC within different realities to be made. Moreover, most of the companies are large, with up to 1800 employees.

3.5.2 RFID technology deployment in FTSC operations

In the world of textile and fashion retailing, the start-up process of a collection assumes, more and more, a strategic role in the competitiveness of a company. When the number of items of apparel begin to increase, it is critical that companies adopt new management paradigms in order to make their processes more agile, thus avoiding not only delays in the reception of items in warehouses, but also bottlenecks in the launching phase of a collection. This is possible through the deployment of RFID technology, which allows the synchronization of supply chain members.

In the companies researched, RFID technology has been deployed in different ways and contexts supporting different kind of operations (Table 3.5). Some of the operations supported include: handling processes; tracking of work-in-progress; receiving operations; shipping operations; tracking products; tracking inventories; monitoring and sorting of merchandise; the tracking, counting of stock and picking of merchandise, tracking containers and shipping; locating products; and store management.

All the case studies deploy RFID to support several operations, with 'tracking products operations' being the most used, followed by 'shipping operations'.

Table 3.3 Case study profile

	Company	Reference	Based	N° Employees
Textile manufacturer	Griva	Bensel *et al.* (2008) Loebbecke (2007)	Italy	70
Clothing manufacturer	Gardeur AG	Wessel (2006) Kernan (2007) Hansen and Giller (2008) Berger (2006)	Germany	1853
Clothing manufacturer and fashion retailer	Lemmi Fashion	Speer (2006)	Germany	80
Fashion retailer	Kaufhof	Loebbecke (2007) Loebbecke and Huyskens (2008)	Germany	19000
Logistics service provider	DHL Solutions Fashion	Ilic (2004)	France	12000

According to the analysis of the five companies, it is possible to infer that companies in the FTSC deploy RFID to support different operations. Therefore, the results support the following proposition:

P_1: Companies in the FTSC can deploy RFID technology to support operations.

3.5.3 Advantages of RFID technology in FTSC operations

Many advantages associated with the deployment of RFID technology were recognized by the case study companies, with the three recognized by almost all the companies being 'better inventory management', 'improvement of read rates' and 'orders' accuracy' (Table 3.6).

Better inventory management was indicated by the clothing manufacturer, Gardeur AG, the clothing manufacturer and fashion retailer, Lemmi Fashion and by the fashion retailer, Kaufhof. Improvement of read rates was identified by the clothing manufacturer, Gardeur AG, the fashion retailer, Kaufhof and also the

Table 3.4 RFID deployment in fashion and textile SC

	Company characterization	Operations in which RFID is deployed	Drivers of RFID deployment	Advantages	RFID system characteristics/ How it works
Griva Textile manufacturer	Griva produces rolls of fabric, much of it used to make flame-retardant curtains. Raw materials are woven into fabric, which is cut and rolled around durable cardboard or plastic cores. Each role of fabric is then mechanically unrolled, treated and re-rolled back onto the cores at 15 separate dye and coating stations in the facility, using a completely automated system. Each roll is identified and tracked at every step in the process, employing RFID hardware.	• Tracking garments • Handling process • Tracking work-in-progress	• To address inefficiencies in its work-in-progress tracking system • To overcome the main disadvantages associated with barcodes such as difficulties in reading barcodes in high temperatures and industrial environments and when the rolls are covered in plastic film (before the RFID deployment, only 70% of the stretch-wrapped rolls were read)	• To improve the reliability of its identification system • Lower labour costs – it contributes to improve productivity • Keeps production lines operating at top speeds • To achieve a 30% return on its investment	Workers apply a label embedded with a passive EPC onto each core. As each roll of fabric exits a treatment station, an interrogator reads the unique Electronic Product Code (EPC) encoded to its core's tag. The RFID middleware removes duplicate reads and sends the tag ID number a production-tracking application, which is part of its enterprise resource-planning platform.
Gardeur AG Clothing manufacturing	The company has its headquarters in Germany. Approximately 90% clothing items are trousers and jeans for men and women, and the	• Receiving operations • Shipping operations	• To get a better view of its supply chain • To reduce shrinkage, to increase its efficiency	• To reduce manual labour • Better account for goods shipped and received	A credit-card-sized passive 13.56 MHz transponder is attached to each item, which has a unique identification (UID) number that cannot be erased.

(Continued)

Table 3.4 Continued

	Company characterization	Operations in which RFID is deployed	Drivers of RFID deployment	Advantages	RFID system characteristics/ How it works
	company supplies them to department stores and up-market boutiques.	• Tracking products	• Many times the company did not know how many garments had arrived at its main warehouse and distribution centre, nor was it able to confirm that all goods shipped from production sites had been delivered to the warehouse	• Better inventory management • More precise fulfilment of manufacturing orders • Less shrinkage • Fewer customer rebates due to a more reliable delivery process • The read rates average is near 100%, allowing managers to confirm that they have actually received all shipped goods	Workers place all garments on hanging carts and wheel them through a 2-m-high gate fitted with an RFID antenna and an interrogator. Here, interrogators read the tags again and send the data to an Open operating software. This software functions as data and device management middleware, and also offers an administration suite that e-mails the IT administrator if an interrogator fails. The Open operating software collects and processes all the data from devices, then formats it to be compatible with Gardline. When goods arrive at the warehouse and distribution centre, the tags are read again as they pass through an RFID portal identical to the one set up.
Lemmi Fashion Clothing manufacturer and fashion retailer	Founded in 1959. Sells about 1 million garments per year to 14 countries. The Distribution Centre is based in Germany.	• Tracking inventories • Tracking products from factory to stores	• The small lots and multiple offerings made inventory tracking and deliveries difficult with barcodes	• Improved speed to market • Better inventory management • Order accuracy	RFID tagging was implemented at the item level. Garments are tagged at the point of production with a one-way RFID hangtag operating at a frequency of 13.56MHz.

	Production Partners in Asia. Leading independent manufacturer of children's wear. Annually, large number of SKUs	• Monitoring and sorting of merchandise • Track, count stock and pick merchandise • Tracking containers • Shipping	• There are inefficiencies in the warehouse • Salespeople sell merchandise that does not exist or cannot be located • There are always garments tucked away in the warehouse • After the goods arrived at its 30 000 m^2 distribution centre (DC) in Germany, the company also needed the ability to quickly track, count, stock and pick merchandise and to better plan its distribution	• Improved quality of the counting	The hangtag contains size, colour and style information that enables the company to track each garment's whereabouts 4 times: 1) in transit from vendor to distribution centre (DC); 2) upon arrival at the DC; 3) as it moves from quality control into its assigned warehouse section; 4) in shipping, where it is reconciled with the purchase order. The RFID hangtags are read by mounted RFID readers, antennae and handheld readers positioned along the supply chain.
Kaufhof Fashion retailer	Galeria Kaufhof, a division of Metro Group, has deployed a fully integrated RFID system in its menswear department.	• Tracking products • Locating products	• High level of stockouts • Necessity to increase cross-selling	• Better control of stock levels on the shop floor and in the warehouse • Increased visibility of the department's tagged items	All the transfer gates in the department store are equipped with FID/EPC technology. RFID readers are generally installed between storage areas and sales floors, as well as on hangers and cash registers.

(*Continued*)

Table 3.4 Continued

	Company characterization	Operations in which RFID is deployed	Drivers of RFID deployment	Advantages	RFID system characteristics/ How it works
		• Store management	• To increase visibility of the department's items • Necessity of controlling stock levels on the shop floor and in the warehouse • Problems with checkout and theft • Problems with locating products	• Significant reductions in stockouts • To identify pieces of clothing that customers are trying in the smart dressing rooms and to display automatically useful information such as price, materials and care instructions Improved reading since the RFID system reads all items on a one-half metre-wide shelf, with a 99% read reliability	The RFID tags are placed on clothing at the department store's regional distribution centre in Neuss-Norf, Germany. Upon arrival at the store in Essen, around 60 RFID readers in the receiving and stock rooms interrogate the tags and track the garments' locations. All the transfer gates in the department store are also equipped with RFID/EPC technology.
DHL Solutions Fashion Logistics service provider	A global logistics service provider for clothing manufacturers and retailers. Each year, 70 million garments pass through the centre on their way to wholesale suppliers.	• Tracking containers • Tracking loads • Inventory monitoring • Searching items in warehouse	• To speed the delivery of products • To enable shipments to be tracked through the supply chain	• To reduce the time needed to unload a truck and check all the goods into the distribution centre • A shipment of 450 tagged garments hanging on a rail	DHL's RFID-enabled room contains three ways to read garment tags. A portal equipped with two antennas reads tags on clothing shipped while hanging on a rail. For garments shipped in boxes, a tunnel reader with three antennas surrounds a conveyor and read tags as the boxes move

with 450 clothing items was scanned in 2 min, while the same task with bar codes takes 8 min • High level of reading reliability	along the conveyor. A mobile reading system with two antennas rides on ceiling-mounted rails and is pushed manually through the room to carry out full inventories. It takes 30 s for this mobile system to take an inventory of 20 000 hanging garments. The mobile system can also be used to search for a specific item. The credit-card-sized smart labels and hangtags were manually attached to boxes and to clothing items, respectively, at customers' warehouses prior to shipping to DHL. Each smart label or hangtag's embedded RFID tag was associated with the same data that was associated with a barcode also printed on that label or hangtag. The data includes the size of the garment, its colour and a reference number. The tagged garments were then shipped to DHL Fashion distribution centre to be inventoried and redirected according to the shipping instructions from their customers.

Table 3.5 **RFID deployment by case studies in their operations**

	Textile manufacturers	Clothing manufacturers		Fashion retailer	Logistics service provider
	Griva	Gardeur AG	Lemmi Fashion	Kaufhof	DHL Solutions Fashion
Tracking products	X	X	X	X	
Handling process	X				
Tracking work-in-progress	X				
Shipping operations		X	X		
Receiving operations		X			
Tracking inventories			X		
Monitoring and sorting of merchandise			X		
Track, count stock and pick merchandise			X		
Tracking containers					X
Locating products				X	
Store management				X	
Tracking loads					X
Inventory monitoring					X
Searching items in warehouse					X

logistics service provider, DHL Solutions Fashion. Orders' accuracy was identified as an important advantage of RFID deployment by the clothing manufacturer, Gardeur AG and the clothing manufacturer and fashion retailer, Lemmi Fashion.

The evidence from Table 3.6 thus supports the second proposition:

P_2: Most of the RFID advantages recognised by the companies in the FTSC are related to inventory management.

3.5.4 Main drivers of RFID technology in FTSC operations

The implementation of RFID technology in the case studies results from the necessity of the companies to overcome identified internal problems and inefficiencies. As can be seen from Table 3.7, each company points out different drivers to the implementation of RFID technology. The main reason for this is that the companies belong to different segments of the FTSC, and thus have different working conditions and face different problems. One driver, though meaning

Table 3.6 Advantages of the RFID technology highlighted by the case studies in the FTSC

	Textile manufacturer	Clothing manufacturer	Clothing manufacturer and fashion retailer	Fashion retailer	Logistics service provider
	Griva	Gardeur AG	Lemmi Fashion	Kaufhof	DHL Solutions Fashion
To improve the reliability of identification system	X				
To improve productivity	X				
To keep production lines operating at top speeds	X				
To achieve a 30% on ROI	X				
To reduce manual labour		X			
Better account for goods shipped and received		X			
Better inventory management		X	X	X	
To minimize shrinkage		X			
Fewer customer rebates		X			
To improve read rates		X		X	X
To improve speed to market		X	X		
Orders accuracy		X	X		
To increase visibility				X	
To reduce stockouts				X	
To make possible cross-selling strategies				X	
To improve loading and unloading productivity					X

Table 3.7 Main drivers of RFID in the case studies

	Textile manufacturer	Clothing manufacturer	Clothing manufacturer and fashion retailer	Fashion retailer	Logistics service provider
	Griva	Gardeur AG	Lemmi Fashion	Kaufhof	DHL Solutions Fashion
To address inefficiencies in working	X				
To overcome difficulties in reading barcodes in high temperatures and industrial environments and when the rolls are covered in plastic film	X				
To improve visibility of supply chain		X			
To know how many garments arrive at the warehouse and distribution centre, and to confirm that all goods shipped from production sites were delivered to the warehouse		X			
Difficulties in inventory tracking and deliveries with barcodes			X		
Inefficiencies in warehouse			X		
Salespeople sell merchandise that does not exist or cannot be located			X		

Table 3.7 Continued

	Textile manufacturer	Clothing manufacturer	Clothing manufacturer and fashion retailer	Fashion retailer	Logistics service provider
	Griva	Gardeur AG	Lemmi Fashion	Kaufhof	DHL Solutions Fashion
There are always garments tucked away in the warehouse			X		
After the goods arrive at distribution centre the company needs to quickly track, count, stock and pick merchandise and to better plan distribution			X		
High level of stockouts			X	X	
Necessity to increase cross-selling				X	
To increase visibility of the department's items				X	
Necessity of controlling stock levels on the shopfloor and in the warehouse				X	
Problems with checkout and theft				X	
Problems with locating products				X	
To speed the delivery of products					X
To enable shipments to be tracked through the supply chain					X

different things to different companies, is common to all: the 'identification of inefficiencies' associated with the development of different operations. For example, inefficiencies are identified as an important driver in the factory, Griva, in warehouses, Lemmi Fashion, in stores, Lemmi Fashion, on the shop floor, Kaufhof, and in the delivery of products, Kaufhof.

Another driver is related to drawbacks associated with barcodes, such as difficulties in inventory tracking and deliveries, Lemmi Fashion, and the difficulty of reading in high temperatures and industrial environment, Griva. Finally, another driver that motivates the deployment of RFID technology among the case study companies is the necessity of accurate counting of products, Gardeur AG and Lemmi Fashion. Thus, the results from Table 3.7 support the third proposition:

> P_3: The main driver to companies in the FTSC to deploy RFID technology is to overcome inefficiencies in operational processes.

3.6 Conclusion

The increased use of RFID technology has been highlighted by several kinds of companies, especially within the fashion and textile world, as a positive factor because of the advantages it brings. The fashion and textile industry has unique requirements associated with short product life-cycle, long production lead time, and high forecasting errors which have motivated companies to deploy RFID. The case studies in this chapter have shown that companies in the FTSC deploy RFID technology to support operations. This clearly supports proposition P_1, with some of the operations supported by this technology and identified by the case study companies being the handling process, tracking work-in-progress, receiving operations, shipping operations, tracking products, tracking inventories, monitoring and sorting of merchandise, counting stock and picking merchandise, tracking containers, locating products, and store management. This result is in line with the literature review (Baars *et al.*, 2009; Bottani, 2009; Butcher, 2007; Miragliotta *et al.*, 2009).

When companies deploy RFID technology they expect to achieve some advantages, usually associated directly with the overcoming of inefficiencies previously identified in the way some operations are performed. The main advantages highlighted by the companies researched are better inventory management, improved read rates and order accuracy. Within these advantages, inventory management is the one recognized by three of the five companies researched. These results make it possible to infer that proposition P_2 is observed in this study, enforcing the same conclusions gathered in the literature review (Jones *et al.*, 2005; Kinsella and Elliot, 2005; Veeramani *et al.*, 2008).

Finally, the main driver that led the case study companies to deploy RFID technology is the identification of inefficiencies in several operations. This same result can be found in Charikleia (2010), Roussos (2008) and Waller *et al.* (2011). Based on this, it can be stated that the proposition P_3 is observed.

This study makes several interesting contributions, from both an academic and managerial perspective. First, it is an empirical investigation of the deployment of RFID technology in FTSC, examining different companies at different SC levels and in different geographical realities. Second, the state-of-the-art of RFID deployment among FTSC companies is explored from the point of view of RFID users, strengthening the validity of the analysis. Moreover, the case-study examination of the main operations supported by RFID technology, its advantages and the main drivers to its introduction form a third important contribution for companies in this industry. However, the limitations of the study should be noted: further research concerning RFID deployment in the FTSC is needed, with more case studies, or a survey to consider a wide sample from around the world. Future researchers should collect large samples of empirical data from the FTSC to enable a deeper analysis of RFID deployment in these kinds of supply chains.

3.7 Acknowledgement

This work was partly supported by Fundação para a Ciência e Tecnologia (FCT) under the project PEst-OE/EEI/LA0008/2013.

3.8 References

Albright, B. (2005), Retailers wrestle with the data end of RFID, *Frontline Solutions*, 6(5), 8–23.

Amato, I., Papadimitriou, T. and Baglieri, E. (2012), The anti-counterfeiting potential of RFID technologies in the fashion supply chain, *17th International Working Seminar on Production Economics*, Innsbruck, Austria.

Atkinson, W. (2004), Tagged: the risks and rewards of RFID technology, *Risk Management*, 51(7), 12–18.

Azuma, N. J., Fernie, J. and Higashi, T. (2009), Market orientation and supply chain management in the fashion industry. In: J. Fernie and L. Sparks (eds), *Logistics and Retail Management: Emerging Issues and New Challenges in the Retail Supply Chain*, London, PA: Kogan Page Ltd., 62–81.

Baars, H., Gille, D., and Strüker, J. (2009), Evaluation of RFID applications for logistics: a framework for identifying, forecasting and assessing benefits, *European Journal of Information Systems*, 18(6), 578–91.

Bensel, P., Gunther, O., Tribowski, C. and Vogeler, S. (2008), Cost-benefit sharing in cross-company RFID applications: A case study approach. In: *ICIS 2008 Proceedings, 29th International Conference on Information Systems*, Paris, 1–17. Available from: aisel.aisnet.org/icis2008/129/ (accessed 13 February 2013).

Berger, D. (2006), RFID-Roll-Out beim Bekleidungshersteller Gardeur AG, in *ident*, 4, pp. 20–1.

Borck, J. (2006), Tuning in to RFID, *InfoWorld*, 28(16), 31–6.

Bottani, E. (2009), The impact of RFID technology on logistics processes of the fashion industry supply chain, *International Journal of RF Technologies: Research and Applications*, 1(4), 225–52.

Bruce, M., Daly, L. and Towers, N. (2004), Lean or agile: A solution for supply chain management in the textiles and clothing industry, *International Journal of Operations & Production Management*, 23(2), 151–70.

Brun, A. and Castelli, C. (2008), Supply chain strategy in the fashion industry: Developing a portfolio model depending on product, retail channel and brand, *International Journal of Production Economics*, 116(2), 169–81.

Butcher, T. (2007), Radio Frequency Identification: An enabler of agile supply chain decision-making, *International Journal of Agile Systems and Management*, 2, 305–20.

Castelli, C. M. and Brun, A. (2010), Alignment of retail channels in the fashion supply chain: An empirical study of Italian fashion retailers, *International Journal of Retail and Distribution Management*, 38(1), 24–44.

Chang, S., Klabjan, D. and Vossen, T. (2010), Optimal radio frequency identification deployment in a supply chain network, *International Journal of Production Economics*, 125(1), 71–83.

Chappell, G., Durdan, D., Gilbert, G., Ginsberg, L., Smith, J. and Tobolski, J. (2003a), *Auto-ID in the Box: The Value of Auto-ID Technology in Retail Stores*, Accenture White Paper.

Chappell, G., Durdan, D., Gilbert, G., Ginsberg, L., Smith, J. and Tobolski, J. (2003b), *Auto-ID on Delivery: The Value of Auto-ID Technology in the Retail Supply Chain*, Accenture White Paper.

Charikleia, L. (2010), *RFID in the Retailing Supply Chain: A Case Study on a Fashion Retailing Industry*, MSc thesis. Available from: hdl.handle.net/2077/22606 (accessed 10 February 2013).

Chen, J. V. and Pfleuger, P. Jr. (2008), RFID in retail: A framework for examining consumers' ethical perceptions, *International Journal of Mobile Communications*, 6(1), 53–66.

Christopher, M., Lowson, R. and Peck, H. (2004), Creating agile supply chains in the fashion industry, *International Journal of Retail and Distribution Management*, 32(8), 367–76.

Coltman, T., Gadh, R. and Michael, K. (2008), RFID and supply chain management: introduction to the special issue, *Journal of Theoretical and Applied Electronic Commerce Research*, 3(1), 3–5.

Cover Pages, 2003, January. METRO group to introduce RFID across process chain. *Cover Pages Press Release*. Available from: xml.coverpages.org/METRO-RFID.html (accessed 7 February 2013).

Cromhout, D., Williams, B. and Hardgrave, B. (2012), *RFID Item-level Quantity Auditing for Apparel Supplier Distribution Centers*, University of Arkansas RFID Research Centre.

Delen, D., Hardgrave, B. and Sharda, R. (2007), RFID for better supply-chain management through enhanced information visibility, *Production and Operations Management*, 16(5), 613–24.

Ellram, L. M. (1996), The use of the case study method in logistics research, *Journal of Business Logistics*, 17(2), 93–138.

Erickson, G. and Kelly, E. (2007), Building competitive advantage with radio frequency identification tags, *Competitiveness Review*, 17(1/2), 37–46.

Finch, P. (2004), Supply chain risk management, *Supply Chain Management: An International Journal*, 9(2), 183–96.

Gaukler, G. M. and Seifert, R. W. (2007), Applications of RFID in supply chains. In: H. Jung, B. Jeong and F. Frank Chen (eds), *Trends in Supply Chain Design and Management: Technologies and Methodologies*, London: Springer, 29–48.

Grant, D. B. and Fernie, J. (2008), Exploring out-of-stock and on-shelf-availability in non-grocery, high street retailing, *International Journal of Retailing and Distribution Management*, 36(8), 661–72.
Hansen, W. and Giller, F. (2008), *RFID for the Optimization of Business Processes*, Cambridge: John Wiley & Sons, Ltd.
Hardgrave, B., Langford, S., Waller, M. and Miller, R. (2008), Measuring the impact of RFID on out of stocks at Wal-Mart, *MIS Quarterly Executive*, 7(4), 181–92.
Hinkka, V., Rahkonen, M., Holmström, J. and Främling, K. (2010), Building supply chain wide tracking systems: Case fashion and book supply chains, *The 16th International Working Seminar on Production Economics*. Innsbruck, Austria. Available from: legacy-tuta.hut.fi/logistics/publications/Innsbruck_2010_SC_wide_tracking.pdf (accessed 13 February 2013).
Ilic, C. (2004), Tracking fashion with RFID, *RFID Journal*, 15 November. Available from: www.rfidjournal.com/article/print/1235 (accessed 13 February 2013).
ILRI (2010), *Item Level RFID Initiative*. Available from: www.vilri.org (accessed 7 February 2013).
Javick, P. (2011), What's the future of item-level RFID? *Apparel, Technology & Business Insight – from concept to consumer*, Available from: apparel.edgl.com/case-studies/What-s-the-Future-of-Item-Level-RFID-75851 (accessed 7 February 2013).
Jick, T. D. (1979), Mixing qualitative and quantitative methods: Triangulation in action, *Administrative Science Quarterly*, 24, 602–11.
Johnston, W. J., Leach, M. P. and Liu, A. H. (1999), Theory testing using case studies in business-to business research, *Industrial Marketing Management*, 28(3), 201–12.
Jones, P., Clarke-Hill, C. and Hillier, D. (2005), The benefits, challenges and impacts of radio frequency identification technology (RFID) for retailers in the UK, *Marketing Intelligence & Planning*, 23(4), 395–403.
Jones, P., Clarke-Hill, C., Hillier, D. and Comfort, D. (2004), Radio frequency identification in retailing and privacy and public policy issues, *Management Research News*, 27(8/9), 46–56.
Kapoor, G., Zhou, W. and Piramuthu, S. (2009), Challenges associated with RFID tag implementations in supply chains, *European Journal of Information Systems*, 18(6), 526–33.
Karkkainen, M. and Holmstrom, J. (2002), Wireless product identification: Enabler for handling efficiency, customisation and information sharing, *Supply Chain Management: An International Journal*, 7(4), 242–52.
Kelly, E. and Scott, P. (2005), RFID tags: Commercial applications vs. privacy rights, *Industrial Management & Data Systems*, 105(5/6), 703–15.
Kernan, A. (2007), Moving to mass markets, *Supply Network*, 39–41.
Khan, M. A., Sharma, M. and Brahmanandha Prabhu, R. (2009), A survey of RFID tags, *International Journal of Recent Trends in Engineering*, 1(4), 68–71.
Kinsella, B. and Elliott, M. (2005), Delivering the goods, *Industrial Engineer*, 37(3), 24–31.
Knill, B. (2002), Pallet tracking leads RFID applications, *Material Handling Management*, 57(1), 8–10.
Kwok, S. K. and Wu, K. K. W. (2009), RFID-based intra-supply chain in textile industry, *Industrial Management & Data Systems*, 109(9), 1166–78.
Lam, J. K. C. and Postle, R. (2006), Textile and apparel supply chain management in Hong Kong, *International Journal of Clothing Science & Technology*, 18(4), 265–77.
Lee, H. and Ozer, O. (2007), Unlocking the value of RFID, *Production and Operations Management*, 16(1), 40–64.

Lo, W., Hong, T. and Jeng, R. (2008), A framework of E-SCM multi-agent systems in the fashion industry, *International Journal of Production Economics*, 114(2), 594–614.

Loebbecke, C. (2004), Modernizing retailing worldwide at the point of sale, *Management Information Systems Quarterly Executive*, 3, 177–87.

Loebbecke, C. (2007), Piloting RFID along the supply chain: A case analysis, *Electronic Markets*, 17, 29–38.

Loebbecke, C. and Huyskens, C. (2008), A competitive perspective on standard-making: Kaufhof's RFID project in fashion retailing, *Electronic Markets*, 18(1), 30–8.

Markides, C. (2007), In search of ambidextrous professors, *Academy of Management Journal*, 50(4), 762–8

McKinnon, A. (2006), Life without trucks: The impact of a temporary disruption of road freight transport on a national economy, *Journal of Business Logistics*, 27(2), 227–32.

Mehrjerdi, Y. (2010), Coupling RFID with supply chain to enhance productivity, *Business Strategy Series*, 11(2), 107–23.

Melski, A., Thoroe, L., Caus, T. and Schumann, M. (2007), Beyond EPC – Insights from multiple RFID case studies on the storage of additional data on tag, *International Conference on Wireless Algorithms, Systems and Applications*, IEEE Computer Society.

Miragliotta, G., Pala, E., Perego, A. and Tumino, A. (2010), RFID technologies to track original products: A general evaluation model and an application to the apparel industry, *12th International Conference on The Modern Information Technology in the Innovation Processes of the Industrial Enterprise (MITIP)*, Aalborg (Denmark), 248–56.

Miragliotta, G., Perego, A. and Tumino, A. (2009), A quantitative model for the introduction of RFID in the fast moving consumer goods supply chain: Are there any profits? *International Journal of Operations & Production Management*, 29(10), 1049–82.

Modrak, V., Knuth, P. and Novak-Marc, J. (2010), Advantages and risks of RFID in business applications, *International Business Management*, 4(1), 28–34.

Mogre, R., Perego, A. and Tumino, A. (2009), RFID-enabled lateral trans-shipments in the fashion apparel supply chain, *Proceedings of the 3rd International Workshop on RFID Technology – Concepts, Applications, Challenges, IWRT 2009*, in conjunction with *ICEIS 2009, International Conference on Enterprise Information Systems*, Milan.

Moon, K. L. and Ngai, E. W. (2008), The adoption of RFID in fashion retailing: A business value-added framework, *Industrial Management & Data Systems*, 108(5), 596–612.

Oppermann, M. (2000), Triangulation – A methodological discussion, *The International Journal of Tourism Research*, 2(2), 141–6.

Perry, C. (1998), Processes of a case study methodology for postgraduate research in marketing, *European Journal of Marketing*, 32(9/10), 785–802.

Roberti, M. (2003), RFID: the cost of being smart. *CIO Insight*, 1(30). Available from: www.cioinsight.com/c/a/Past-News/RFID-The-Cost-of-Being-Smart/ (accessed 7 February 2013).

Roussos, G. (2008), Computing with RFID: Drivers, technology and implications, *Advances in Computers*, 73, 161–217.

Rowley, J. (2002), Using case studies in research, *Management Research News*, 25(1), 16–27.

Sarac, A., Absi, N. and Dauzère-Pérès, S. (2010), A literature review on the impact of RFID technologies on supply chain management, *International Journal of Production Economics*, 128(1), 77–95.

Seuring, S. (2004), Integrated chain management and supply chain management? comparative analysis and illustrative cases, *Journal of Cleaner Production*, 12(8/10), 1059–71.

Shapiro, D. L., Kirkman, B. L. and Courtney, H. G. (2007), Perceived causes and solutions of the translation problem in management research, *Academy of Management Journal*, 50(2), 249–66.

Singleton, R. A. and Straits, B. C. (1999), *Approaches to Social Research*, New York: Oxford University Press.

So, S. and Liu, J. (2006), Securing RFID applications: issues, methods, and controls, *Information Systems Security*, 15(4), 43–56.

So, S. and Sun, H. (2010), Creating ambient intelligent space in downstream apparel supply chain with radio frequency identification technology from lean services perspective, *International Journal of Services Sciences*, 3(2/3), 133–57.

Speer, J. K. (2006), Lemmi fashion making (13.56) waves, *Apparel Magazine*, 47(6), 22–4.

Stuart, I., Deckert, P., McCutcheon, D. and Kunst, R. (1998), Case study: A leveraged learning network, *Sloan Management Review*, 39(4), 81–93.

Ustundag, A. (ed.) (2013), *The Value of RFID, Benefits vs. Costs*, London: Springer Verlag.

Veeramani, D., Tang, J. and Gutierrez, A. (2008), A framework for assessing the value of RFID implementation by Tier-one suppliers to major retailers, *Journal of Theoretical and Applied Electronic Commerce Research*, 3(1), 55–70.

Visich, J. K., Li, S., Khumawala, B. M. and Reyes, P. M. (2009), Empirical evidence of RFID impacts on supply chain performance, *International Journal of Operations & Production Management*, 29(12), 1290–315.

Waller, M. A., Cromhout, D. and Patton, J. B. (2011), *An Empirical Study of Potential Uses of RFID in the Apparel Retail Supply Chain*, University of Arkansas RFID Research Centre.

Walton, C. A. (1983), *Portable radio frequency emitting identifier*, US patent 4384288, 17 May 1983. Available from: www.google.com/patents?vid=4384288 (accessed 7 February 2013).

Wessel, R. (2006), Clothing manufacturer invests its ROI in RFID, *RFID Journal*. Available from: http://www.rf-it-solutions.com/uploads/tx_easydownloadlist/200608_gardeur.pdf.

Williams, B. and Waller, M. (2010), Creating order forecasts: Point-of-sale or order history? *Journal of Business Logistics*, 31(2), 231–51.

Williams, B. and Waller, M. (2011), Estimating a retailer's base stock level: An optimal distribution centre order forecast policy, *Journal of the Operational Research Society*, 62(4), 662–6.

Witt, C. (2006), Real-world ROI for RFID, *Material Handling Management*, 61(5), 26–31.

Yin, R. K. (2002), *Case Study Research: Design and Methods*, 3rd edition, *Applied Social Research Methods Series*, vol. 5, Thousand Oaks, CA: Sage Publications, Inc.

Zhou, W. and Piramuthu, S. (2012), Manufacturing with item-level RFID information: From macro- to micro-quality control, *International Journal of Production Economics*, 135(2), 929–38.

4

The role of radio frequency identification (RFID) technologies in improving garment manufacturing operations[1]

E. W. T. NGAI, D. C. K. CHAU, J. K. L. POON, A. Y. M. CHAN, B. C. M. CHAN and W. W. S. WU,
The Hong Kong Polytechnic University, Hong Kong

DOI: 10.1533/9780857098115.70

Abstract: A growing number of organisations are using radio frequency identification (RFID) technology to improve their manufacturing processes. A case study approach is used to explore the development of an RFID-based manufacturing process management system in a garment factory in China. The results indicate that both technology push and need pull factors influence the garment factory's intention to adopt RFID technology. We identify eight factors for the successful implementation of an RFID-based manufacturing process management system, namely, vendor selection, organisational motivation, cost/benefit evaluation, top management support, user involvement, the extent of progress supervision, staff competence and training, and policy, structure and operating process compatibility. This in-depth case study details the organisation's experience and identifies the challenges it faces and important issues in the development and implementation of the system. Implications for academics and practitioners are discussed.

Key words: RFID, manufacturing process management system, success factors, case study, garment industry.

4.1 Introduction

Advances in Information Technology (IT) present many opportunities for companies, including the garment and textile industry, to enhance and transform their work processes and business relations (Sambamurthy and Zmud, 1999; Subramanian, 1990). IT is proven to have a potential impact on organisational operational efficiency and competitive advantage (Lengnick-Hall *et al.*, 2004; Melville *et al.*, 2004). In recent decades, radio frequency identification (RFID) has attracted the attention of the manufacturing industry and other businesses, where it is increasingly being used (Ngai *et al.*, 2008a). Even industries with lower profit margins, such as

1 This chapter has been published as: E. W. T. Ngai, D. C. K. Chau, J. K. L. Poon, A. Y. M. Chan, B. C. M. Chan and W. W. S. Wu (2012), Implementing an RFID-based manufacturing process management system: Lessons learned and success factors, *Journal of Engineering and Technology Management*, 29(1), 112–30. Reproduced with permission from Elsevier.

the garment manufacturing industry, are implementing this technology, as the costs of RFID tags come down and their benefits are recognised (Ngai *et al.*, 2007). RFID facilitates communications with identified objects using radio waves, meaning that these objects can be recognised, tracked and traced (Jones *et al.*, 2005). It is believed that RFID can help organisations to track assets and monitor key performance indicators more accurately, make manufacturing operations more visible and allow decisions based on real-time information (Loebbecke *et al.*, 2006).

In the textile and garment industry, while RFID technology has been applied in stocktaking operations, strengthening customer relationships, facilitating promotional activities and allowing retailers to allocate resources more effectively (Moon and Ngai, 2008), the use of this technology in garment manufacturing has not been fully investigated. In traditional garment factory manufacturing processes, after the manual tracking system has sent material to the production shop floor, it enters a 'black hole', meaning that production is not visible to management until the finished product emerges. Most garment manufacturers still use 'paper tickets' to manage their operations. Each bundle is tied with a paper ticket that has a barcode label for the different operational steps. Once sewers have completed their process, they tear off their part of this paper ticket. At the end of the day, the supervisor collects all of the tickets, and a clerk enters the sewers' staff identifiers and scans their tickets.

Operational information is thus collected and input into the system at the end of one day's operations. With the aim of improving the garment manufacturing process with RFID technology and motivating manufacturers to derive its full benefits, this chapter presents a case study that demonstrates how an RFID-based manufacturing process management system can collect real-time information and increase operational efficiency. The research is based on a single in-depth study that is mostly used for exploration. The experience of and lessons learnt (LL) from this study will add to the body of knowledge on Operations Management and Information Systems (IS).

In Section 4.2, we review the business value of RFID in the garment industry and consider the 'technology-push' (TP) and 'need-pull' (NP) factors that underlie the adoption decision. In Section 4.3, we present our experiences of the actual implementation of an RFID-based manufacturing process management system in a garment factory. In Section 4.4, we summarise the motivating factors behind the adoption of RFID technology. In Section 4.5, we present the critical success factors, as derived from the case study experiences. In Section 4.6, we present the LL. Finally, we conclude the chapter and provide directions for further study.

4.2 The business value of RFID technology in garment manufacturing operations

RFID is a wireless technology that uses transmitted radio signals to tag, recognise, track and trace the movement of an object automatically (Loebbecke *et al.*, 2006; Moon and Ngai, 2008). It is one type of auto identification technology, which also includes magnetic strips and the ubiquitous barcodes (Delen *et al.*, 2007). Basically, RFID infrastructures are composed of three principal components (Shepard, 2005):

1. *a transponder*: is a reusable tag with a tiny computer chip to identify an object;
2. *a reader*: serves as a communication channel between the tag and the RFID-based IS; and
3. *a data collection application*: comprises a software solution and an antenna.

According to Ngai *et al.* (2008a), the applications of RFID technology are diverse and have been applied in at least 14 different industries, including retailing, library services, and logistics and supply chain management. Despite the growing recognition of this technology, the garment manufacturing industry, a labour-intensive industry that is difficult to automate, has not yet widely adopted it. Most organisations in this industry use manual or partial barcoding systems to keep track of the production line.

Despite the potential business value of RFID technology, the current cost of RFID tags is of great concern to the garment manufacturing industry in considering its adoption. Although the price of these tags is dropping (now about US$1), the cost is still relatively high for this industry. The major portion of the profits in the fashion supply chain goes to the retailer. The original equipment manufacturing (OEM) garment industry generally has only a 10% profit margin. Thus, the relatively high cost of RFID tags may act as a barrier to its adoption. However, with the rapid advancement of technology, the price of these tags has dropped tremendously, with the latest developments promising tags that cost as little as 10 to 20 pence each for a small batch or a penny each for volumes in the hundreds of millions (Little and Arnst, 2002; Prasad, 2007).

In recent years, the performance gap between expected and actual production line efficiency and effectiveness has become larger. As garment manufacturing is a labour-intensive industry, output quality and under-/over-production problems occur frequently. Manufacturers require real-time information on the production line to better manage operations and solve the operational problems that occur before it is too late. However, the operating costs of China's garment manufacturing industry are increasing rapidly, due to stricter product quality requirements, increasing unit labour costs and stricter regulations on labour rights and working hours. This industry needs to enhance the efficiency and effectiveness of its production system so as to remain competitive in the market. Moreover, we foresee that the lower end of the fashion supply chain, retailers, will apply increasing pressure on garment manufacturers to adopt RFID technology for the better management of the logistics process and the retail supply chain. Wal-Mart, one of the world's largest retailers, demands that important suppliers include RFID tags with their shipments (Sellitto *et al.*, 2007). We predict that this trend will spread across industries, especially to those with a shorter product life-span and greater demand for individual customisation.

We conducted a literature search on the following well-known electronic databases, in order to report the status of extant literature related to the success factors and the LL from the implementation of RFID: ABI/INFORM, Academic Search Premier, Emerald Fulltext, Science Direct, and IEEE/IEE Electronic Library. The descriptors 'RFID', 'success factors' and 'lesson learned' were used in the literature search. These descriptors were selected because we wanted to know how well the success factors of the LL from the implementation of RFID system were studied. The following presents a brief description of how these topics were addressed in studies published in academic journals.

Leimeister *et al.* (2009) developed a model that shows how the intention to invest in RFID technology by Chief Information Officers (CIOs) from Germany and Italy is affected by different factors. The research also showed that cultural differences affect how CIOs perceive the value of RFID. Eight success factors for the successful implementation of RFID system in an aircraft engineering company were proposed in a study conducted by Ngai *et al.* (2007). Lessons concerning the implementation of RFID at strategic, management and operational levels were summarized in order to provide guidelines for practitioners interested in implementing RFID. Attaran (2007) proposed a number of success factors that influence RFID adoption and the barriers deterring businesses from adopting RFID technology. Both the success factors and barriers were identified from past literature.

Delen *et al.* (2007) proposed three critical hurdles to the adoption of RFID system based on the RFID collected by a retailer. All the hurdles resulted from the common problems of RFID data. In a study regarding the implementation of an RFID system for process management of structural steel works in buildings (Chin *et al.*, 2008), three factors that affect the implementation of the system were proposed: the manufacturer's level of computerization, the balance of benefits, and the standard and compatibility of RFID technology. Konomi and Roussos (2007) proposed several important considerations in the implementation of RFID based on the development and experience of two ubiquitous computing projects. Ngai *et al.* (2008b) suggested several lessons from their experiences in implementing an RFID-based sushi management system, which can be used as guidelines for future projects. The benefits and challenges of the implementation of the system were enumerated.

In a study conducted by Janz *et al.* (2005), seven 'best practices' in implementing an RFID system were proposed, based on their experiences in the implementation of an RFID-based patient management system in a medical centre. Three concerns regarding data issue concerning RFID that may hinder organisations from implementing an RFID system successfully were also proposed. Some benefits of implementing RFID systems, based on an RFID literature review, were listed in the research conducted by Sellitto *et al.* (2007), which aimed to provide guidelines to potential users of RFID systems. In contrast with other studies that assume businesses can independently make decisions regarding the use of RFID technology, Lai *et al.* (2006) used the institutional isomorphic perspective to explain the adoption of RFID technology for supply chain management.

As pointed out by Ngai *et al.* (2008a), the critical success factors of RFID adoption merit further investigation. However, previous studies concerning the success factors of RFID adoption and the LL from RFID implementation have not focused on the adoption of RFID technology in the garment manufacturing industry. Therefore, the critical success factors of RFID adoption in this industry have not yet been thoroughly studied. Surprisingly, there has also been little effort in providing practitioners guidance for RFID system implementation (Ngai *et al.*, 2010).

We adopt the TP and NP perspective to investigate the underlying motivations behind the garment manufacturing industry's investment in RFID technology. The concepts of TP and NP are borrowed from the engineering/R&D management discipline to study the key factors in the adoption decision of a new technology (Shih *et al.*, 2008). They represent two schools of thought that support two different arguments. TP factors result from a recognition of the business value that the new technology could bring to the industry, whereas NP factors result from the user's needs (Chau and Tam, 2000). From this perspective, the decision to adopt RFID in the garment industry could stem from recognition of the business value of this technology, a performance gap, or both. Therefore, TP and NP factors serve as the basic framework in our case study for investigating the motivation to adopt the new system.

4.3 Using RFID technology in garment manufacture: a case study

A case study can be used for many purposes – exploring (new areas), describing (complex events or interventions) and explaining (complex phenomena) (Kohn, 1997; Lichtenthaler, 2004). This study describes a unique case investigation providing the details of the design and development of the RFID-based garment manufacturing IS (exploring a new area), which explains a complex phenomenon in garment manufacturing processes. An in-depth case study of the development of an RFID-based garment manufacturing process management system in a garment factory in China was conducted to explore and identify the underlying motivations and implementation issues. As a research method, case studies, certainly single-case studies, score low on the generalisability of findings (Akkermans and Van Helden, 2002). However, the richness of their data means they lend themselves well to developing an understanding of the unique characteristics of RFID-based system development within the garment manufacturing industry. We collected information for the case study by referring to internal organisational documents and interviewing key people who have participated in the implementation of the RFID system. One of the authors conducted interviews with the users and members of the management group in preparation for the case analysis.

The case study object ('the company' hereafter) is one of the largest garment manufacturers headquartered in Hong Kong and produces apparel products for internationally renowned brand names for a number of US and European retailers. The company is listed on the main board of the Stock Exchange of Hong Kong. Its production facilities are spread over three continents: China and Indonesia in Asia, El Salvador in Central America and Lesotho in sub-Saharan Africa. The company is principally engaged in the manufacture of knit tops for men, women and children, according to specifications and designs supplied by customers and sold under the customers' own brand names. In 2007, the group set up a manufacturing enterprise in Guangdong Province in China that mainly produces and sells knitted products.

The company was reliant on the traditional manual and partial barcoding system (which scans the bar on the paperwork that accompanies parts of the product) to manage its production operations. Operational information was collected and input into the system at the end of each day's operations. Such a system has certain performance deficiencies and disadvantages. The most obvious disadvantage is the lack of real-time data. It is thus difficult for the production line manager to keep track of the production process and promptly discover operational problems, such as sewing errors or a deviation from the design.

However, many external factors, such as China's serious inflation, RMB inflation, the reduction of export rebates and an increase in the minimum labour wage, have affected the factory's profitability. The company thus needed to enhance production quality and efficiency to remain competitive. Also, its competitors in China had begun to adopt RFID technology. At the same time, the company has had pressure from clothing designers to manufacture a variety of styles for each season, with increased demand for shorter lead times and more timely and fashionable products from retailers. These pressures led it to look for a means of monitoring the manufacturing process in real time with the capability to control each production step and react more quickly to order changes. More specifically, the company in this study faced the following problems:

- The production workers (the sewers) in the company liked to hold up some of the 'job tickets' to increase their overtime pay, which led to inaccurate estimations of workers' daily productivity.
- It was difficult for the company to quantify production inflows and outflows and the actual productivity of manpower and machinery.
- There was a lack of real-time data to identify bottlenecks in the production process and balance the production of different production lines.

Therefore, the company needed greater production flow visibility to gain a better understanding of what was going on in the production flow and to have better coordination of employees on the factory floor. Therefore, through this case study, we aim to illustrate how RFID technology can help to enhance the operational visibility, efficiency and effectiveness of the garment manufacturing process.

With regard to the potential RFID capability for and performance gap in the garment manufacturing process, this study aims to investigate the motivations underlying the RFID adoption decision in the garment manufacturing industry and the success factors of an RFID-based manufacturing process management system. More specifically, the study aims to answer the following questions:

- What are the motivating factors, from the TP and NP perspectives, that drive a garment manufacturer to adopt RFID technology?

- What is the actual business value of an RFID-based manufacturing process management system for a garment manufacturing factory?
- What are the success factors in implementing an RFID-based manufacturing process management system?

4.4 Implementing an RFID-based manufacturing process management system

In 2007, the company began to implement an RFID-based manufacturing process management system. This system, which is provided by a Malaysia-based company (the 'vendor' hereafter), focuses on the key production processes – cutting, sewing, finishing, ironing, quality control (QC) and rework (Fig. 4.1).

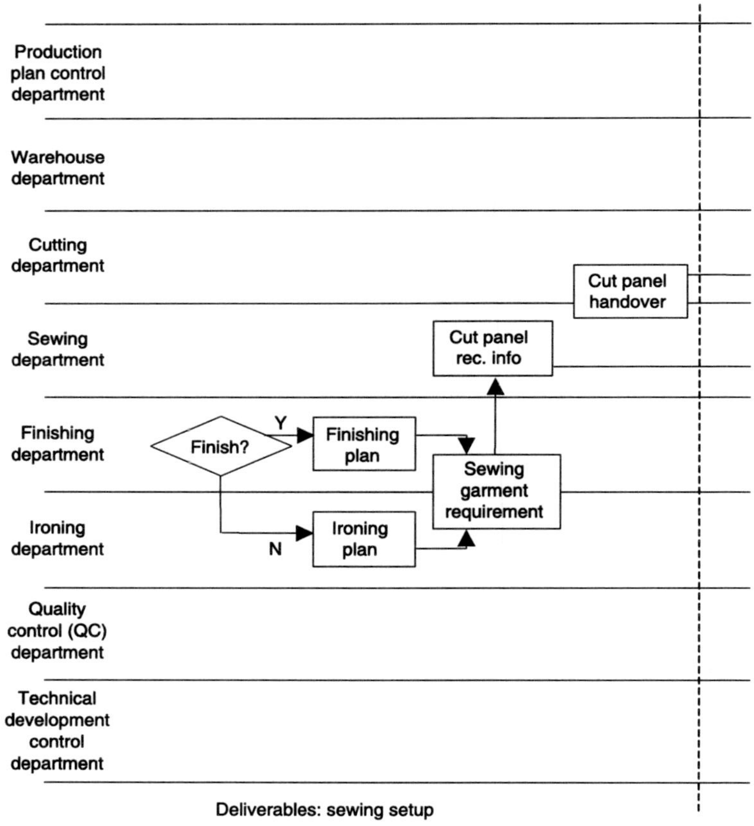

4.1 Workflow of garment production process in the company.

The major components of the RFID-based manufacturing process management system include:

- *An RFID token*: a low-frequency (13.56 MHz), passive tag with a unique ID that is associated with a bundle of cut-raw materials (i.e. sleeves, cuffs and hoods) that are to be used to fulfil each order.
- *RFID readers*: installed in the cutting department next to each sewing machine and QC table (these readers are of two types: that attached to the sewing machine is for reading data only and features simple buttons for enquiries about previously scanned data, whereas that installed in the QC table has a keyboard on which the inspector can input defect codes (e.g. 0 – low quality, 1 – missing step, 2 – fabric problem, 3 – size problem) after a particular piece of the garment has been scanned) and has the memory to store data of up to 3000 records in case of a power shortage.

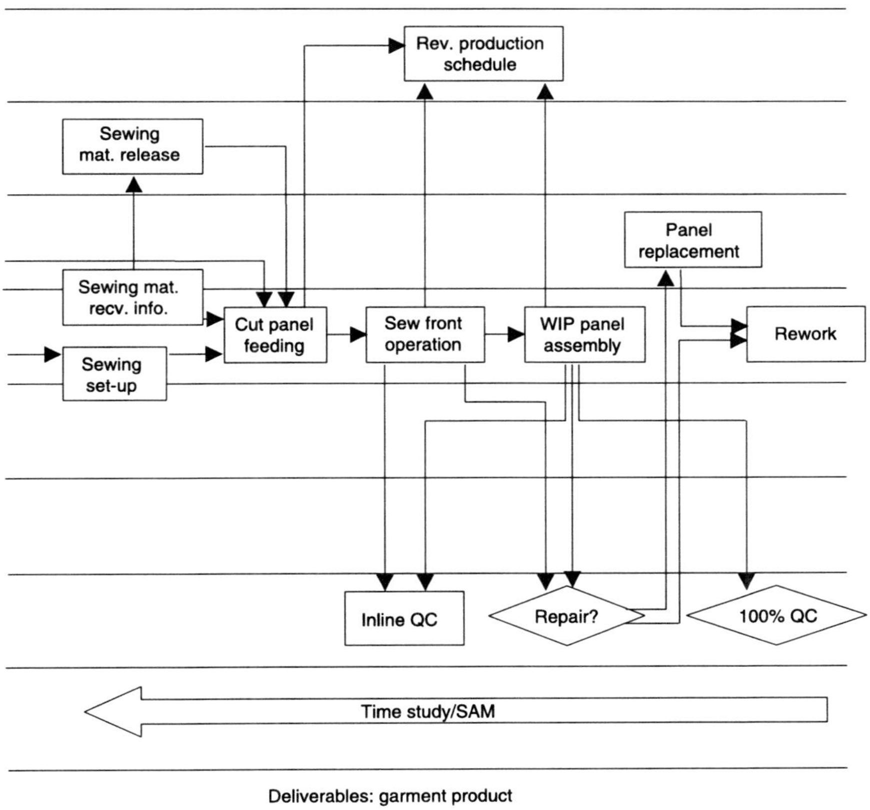

- *A PC workstation*: connects 40 sets of RFID readers in the business structure (Fig. 4.2) and on which data are collected before being uploaded to the computer server.
- *An Ethernet system*: used to transmit data from the PC station to the server.

The system runs in Client-Server technology. The backend was developed by Interbase and the client side by Dephi. The IS supports the generation of several operational reports, for example:

- A daily performance report that shows the performance of a particular sewing group on a daily basis.
- A graphical representation of the efficiency of different sewers on a particular date.

4.4.1 System implementation: Stage I Trial run (August to December 2007)

In the trial-run stage, some of the sewing lines (8 lines) had the RFID-based manufacturing process management system installed. Barcoding (the original process management technology) and the RFID technology operated in parallel. The data generated from the barcoding system could be used to verify the data from the new system. The parallel running of these two types of technology allowed us to test the compatibility and usability of the new system. At the same time, it allowed the sewers and production line manager to adapt to the new system and payroll policies. The trial run also proved to the sewers that the new payroll policy (a bonus calculated on the basis of actual standard sewing minutes (SAM) and an estimated SAM with minimum pay protection) that was implemented together with the new system did not decrease their pay cheques when compared with the barcoding system. This helped the system to gain user acceptance.

4.4.2 System implementation: Stage II Full implementation (January to May 2008)

After the trial-run period, all of the sewing lines had the RFID-based manufacturing management system installed, and the original barcoding system was phased out. The sewers' pay is now calculated on the basis of the information generated by the new RFID-based system. The production line managers use the system reports, which are generated using real-time data, to monitor operations, and continuous training is carried out due to the high staff turnover rate. Also, a new department, the support team (Fig. 4.3), was set up, and staff from the Industrial Engineering Department (IED) were involved in managing the data collected. The IED is a three-member team responsible for analysing the data and measuring pre- and post-system performance to monitor the line-balance.

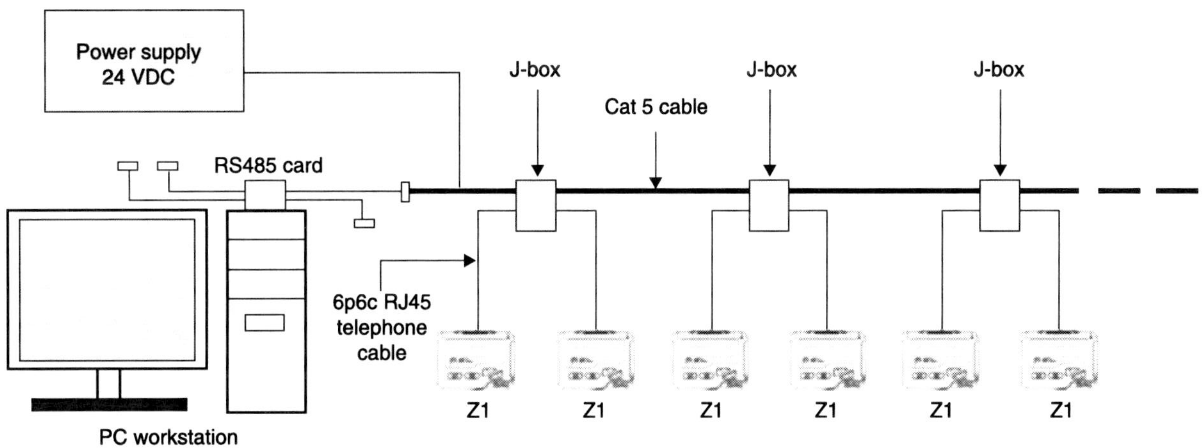

4.2 Set-up of Smart Term Z1 network and cabling connection.

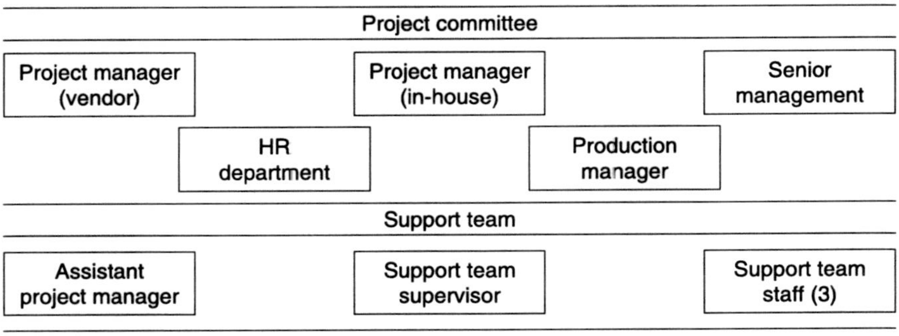

4.3 Project team structure.

The results of the study show that the actual time taken for individual steps in the manufacturing process can be recorded, and factory floor managers can easily obtain better line-balancing through the availability of real-time data from the RFID-based manufacturing process management system. Without such real-time information, factory managers can only guess, based on their experience, the time taken to complete each step of the work. With real-time information, they know exactly how much time is needed for each step of the manufacturing process, thus allowing them to assign sewers accordingly and respond more quickly to production needs. The performance of individual sewers can also be monitored using the real-time data provided by the system. Figure 4.4 shows a comparison of sewer efficiency in the factory on a particular date in a factory. The employee efficiency report provides daily production performance data of each employee. Companies that have multiple employees (e.g. sewers) performing the same operation on a given part number, thus can track the employee's production efficiency compared to production standards for a particular part/operation and to other employees' (sewer) performance.

4.4.3 System implementation: Stage III System application in operational enhancement (June 2008 onwards)

With the new system, operational problems (late arrival of fabric, raw material shortage, mistakes in coordination, etc.) can be highlighted in the operational reports that are discussed in the weekly production meeting. New system modules were added in step-by-step, for example, the Industrial Engineering Execution System (IEES), which employs historical data to calculate the standard time for sewing processes. For the general factory floor user, more advanced user training is offered to teach them how to use the system to locate lost time on the production floor and how to use it to handle colour-matching problems. Shop floor managers are trained to use the system to identify production bottlenecks and other

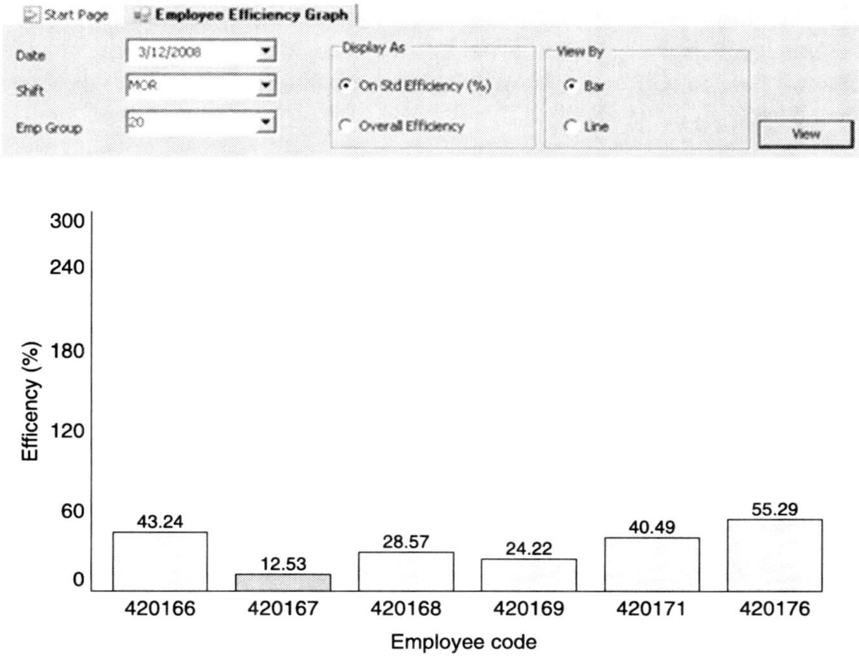

4.4 Employee efficiency in production.

line-balancing problems. At present, there are around 600 sewers using the RFID-based manufacturing process management system, and 100 other users from the merchandising, production planning, industrial engineering and production departments.

4.5 Assessing the effectiveness of RFID technology: data collection and analysis

During the development and implementation of the RFID-based manufacturing process management system, information was collected from different sources to answer our research questions with regard to the motivation for the adoption decision, the system business value, the success factors and critical implementation issues. Empirical materials were collected by interviewing key actors and studying meeting minutes and project plans during the system's development and implementation process.

As pointed out by Stake (1995), interviews should adequately cover the various actors and management levels involved in the system development and implementation processes being studied. Therefore, in this study, we selected the project manager, representatives of top management, a user support team member, different levels of users and the software vendor as our sources of information. Reliability was achieved by using a standardised interview protocol for each case, in which a series of 'how' and 'what' types of research questions were asked.

Face-to-face interviews, each lasting approximately 45 to 60 minutes, were conducted. During the interview, one researcher was in charge of the interview, while another was tasked with taking down notes and observing. Extensive field notes were taken during the interview, as the interview guide was designed to accommodate note-taking. At the end of the interviews, the two researchers discussed them and generated a descriptive report, which were then returned to the interviewees for comments. At least two interviewers took part in each interview. To ensure quality and integrity, transcript preparation comprised two stages, compilation and review. In the former, the researcher transcribed the interview notes taken during each interview session. In the latter, the integrity and correctness of the content of the transcript were verified by another researcher who had been involved in the interview, but who had not been the one to compile the transcript. This approach was used to ensure quality, accuracy and correctness.

In addition to the interviews, information was collected from documents, such as meeting minutes and in-house project documents. The raw data from the interview transcripts, field notes and minutes were stored in a case-study database for further reference and to check for accuracy. To reduce the likelihood of systematic bias, the interviewees' accounts were first analysed separately by two researchers, who then discussed the results. This follows Denzin's (1978) four basic types of triangulation (i.e. data, method, investigator and theoretical). Triangulation through multiple data sources and methods of information collection and analyses increases the credibility of a study (Green *et al.*, 2002; Yin, 1994). The data and information collected from multiple sources of data using different methods were analysed to identify motivation factors, system business value, success factors and other critical implementation issues.

4.6 The business value of the RFID-based manufacturing process management system

The RFID-based manufacturing process management system has led to various performance improvements in the company. In general, it has successfully enhanced production line efficiency and effectiveness by increasing production line visibility and reducing machine down-time, non-working time for sewers, the defect rate, late delivery and the amount of overtime. These operational improvements have helped to decrease operational costs and enhance profitability.

In addition to this operational improvement, the RFID-based system has also helped to uncover previously hidden problems, such as poor coordination among

Table 4.1 List of benefits of RFID-based manufacturing process management system

Tangible benefits
1. *Machine down-time* reduced by 30% compared with last year's figures
2. *Lost time* (non-working time) reduced by 50% compared with last year's figures
3. *Defect rate* reduced by 30% compared with last year's figures
4. *Efficiency* increased by 50% compared with last year's figures
5. *On-time delivery* rate increased to 90%
6. *Amount of over-time working hours* among sewers reduced by 10% on average
Intangible benefits
1. Uncovered previously hidden problems, such as: • Co-ordination problems among departments • Working efficiency of maintenance department • Sewers' skill set deficiency and incompleteness
2. Provides more accurate data for the calculation of SAM, which can then be used for cost estimation and line-balancing
3. Provides real-time production data for the closer monitoring of production defects, downtime, efficiency, production line visibility and other issues, such as process reengineering
4. Has built a culture of open discussion and scientific management, which is linked to the concept of continuous improvement
5. Improved staff motivation by creating an image of a company that is willing to invest in helping employees to enhance work efficiency and obtain a larger pay cheque and bonus

departments, create a culture of continuous improvement, and provide more accurate data for performance evaluation and payroll calculation. Although these positive effects do not contribute to organisational profitability directly, they have helped the organisation to improve operations and create a performance-enhancing organisational culture and boost staff motivation. Therefore, the adoption of an RFID-based manufacturing process management system contributes to organisational performance and profitability in both tangible and intangible ways. The benefits are summarised in Table 4.1.

Several interviews and analyses of project planning and meeting minutes were conducted to identify the factors that motivated the garment factory to adopt the new RFID-based manufacturing process management system. This analysis was carried out according to the TP and NP perspectives. The motivating factors were then identified from the case study and summarised as shown in Table 4.2. TP factors are those that result from a recognition of the RFID benefits to the organisation. In this case study, we found that the following factors were influential in the decision to adopt the technology:

- relative advantage
- compatibility
- complexity
- extendibility
- cost.

Relative advantage has been identified as the key influential factor in the adoption of the new technology (Brown and Russell, 2007; Sharma and Citurs, 2005). All of the interviewees, except for the sewer, recognised the relative advantages of the RFID system. However, those in different positions perceived different advantages. For example, the in-house project manager from the company's Hong Kong headquarters perceived that the system served as a tool to allow senior management to monitor the factory status. The factory manager perceived that the system served as a tool to improve efficiency and discover production problems. The supervisor considered it to be a staff-monitoring tool. The recognition of the advantages of the new system by employees at different organisational levels affected the organisation's adoption decision.

Compatibility is defined as the degree to which a technology is perceived to be consistent with an organisation's strategic objectives, infrastructure, practices and needs (Brown and Russell, 2007). In this case study, the manufacturing company did not want its factory sewers to know the piece rate and quantity of output when the new system was adopted. The standard feature of the RFID reader allowed the company to hide this kind of information from the sewers. Also, the company expected the new system to be compatible with the current information it used. These two requirements were met by the vendors of different IS, which enhanced the company's intention to adopt the new technology. Similar to the results of previous studies (Brown and Russell, 2007; Moon and Ngai, 2008), the adoption decision is affected by a technology's compatibility with organisational requirements.

In the garment manufacturing industry, the primary labourers (the sewers) do not have a high-level educational background. It is thus difficult for them to learn

Table 4.2 Technology push and need pull factors for adoption of an RFID-based garment manufacturing IS

Technology push	Need pull
• Relative advantage	• Competitors
• Compatibility	• Customer
• Complexity	
• Extendibility	
• Cost	

to use new technology that is complex and technical. Thus, the RFID-based manufacturing process management system, which is very easy to learn and use, suited the needs of the sewers and enhanced company confidence in the degree of success that could be expected in its adoption. Therefore, in line with the findings of Brown and Russell (2007), the complexity of technology, particularly its usability and acceptance by potential users, affects the adoption intention.

The extendibility of the new technology to support other operational requirements also affected the adoption decision. As noted by the vendor's project manager, the new system can be plugged in with other sub-modules, for example IEES, which is a centralised repository solution designed to optimise the process of standard time development and management of the garment production process through an industrial engineering approach.

With IEES, the following can be achieved:

- labour cost and grading analysis;
- standardising and synchronising technical motions;
- creating and tracking detailed work methods;
- calculating and storing labour standards;
- time-standard analysis;
- working environment evaluation and improvement;
- data sharing between branches;
- 'plan vs. actual' analysis;
- benchmarking of standards;
- scientific estimation approach;
- creating and tracking cost estimation.

Although cost is believed to be a major consideration in the technology adoption decision (Brown and Russell, 2007), we did not find it to be the major concern in this case study. Other factors, such as relative advantages, customer push and vendor relationships, were more dominant. As noted by the in-house project manager, 'Cost is not the main concern in this case. Other factors (vendor relationship, support from the vendor) were more dominant in this case. We picked this vendor even though it had the highest price among the others.'

The factors related to NP result from the need of the garment manufacturing industry to adopt the new technology. In this study, we identified competitor and customer pull as factors that affected this adoption decision in this industry. The operating costs for China's garment manufacturing industry have increased rapidly, due to stricter product quality requirements, increased unit labour costs and stricter regulations for labour rights and working hours. This industry needs to enhance the efficiency and effectiveness of its production system to remain competitive in the market. Therefore, a competitor's decision about new technology adoption would affect an organisation's own decision. As noted by the in-house project manager, 'Some of our key competitors have started to use RFID, which pushed us to catch up with them.'

Customer pressure on price is another driver forcing organisations to lower production costs by means of new technology. As the factory manager stated:

> Owing to the economic downturn of the US, there is high price pressure from customers. Also, because of appreciation of the RMB, the cost increase in materials and payroll, our production cost is getting higher and higher. The only way we can survive is to increase our efficiency.

In addition to these TP and NP factors, the case analysis found that the vendor relationship and level of support, the attitude of top management, employee competence and past experience all affect the new technology adoption decision. Through the information collected, we discovered that the relationship with and level of support from the vendor affect the adoption decision. As noted by the in-house project manager, 'The vendor is a Malaysia-based company, and our chairman is Indonesian Chinese. The senior management of the two companies share a similar culture, and they know each other well.' The factory manager noted that: 'The support of the vendor is important. It is not just a one-off support for installation. It is a long-term partnership.' These statements show that the relationship with and support from the vendor also affect the intention to adopt new technology.

Similar to previous studies of new technology adoption intention (Loebbecke *et al.*, 2006; Moon and Ngai, 2008; Nystrom *et al.*, 2002; Teo *et al.*, 2004), the attitude and support of top management have proved to be important in the adoption and diffusion of new technologies. As the factory manager said, 'The determination of senior management and the cooperation of different departments are important to the success of the project.'

Staff capability to adapt to new technology and user acceptance are also important in the adoption decision. As the vendor's project manager said, 'The internal staff was willing to learn, and picked it up very fast.' Also, some of the other interviewees noted that 'getting staff acceptance' is important. This was especially true in this case, as the implementation of the new system would affect the sewers' pay.

Past experience of the adoption of new technology is helpful in encouraging an organisation to adopt it in its subsidiaries. In this case, the company had already implemented the system successfully in its factory in Indonesia. This past experience served as a model of the benefits of adoption and helped to secure the support of top management and potential users.

4.7 Factors in successful implementation of an RFID-based manufacturing process management system

Based on our experiences at the company, we have identified eight success factors that we consider to be relevant to the successful implementation of the RFID-based manufacturing process management system. These are summarised in Table 4.3, followed by further elaboration for each factor.

4.7.1 Vendor selection

We found that cost was not the major concern in the selection of vendors. Instead, the vendor's experience in implementing the system in the operational environment of the adopting company, trust between the vendor and the adopting company, and the extent of post-installation vendor follow-up and evaluation were more

Table 4.3 List of success factors for RFID-based system implementation

Vendor selection	Vendor's experience, trust, extents of follow-up work and evaluations after installation are important. A project manager (in-house) said: 'The support of the vendor is important. It is not just a one-off support for installation. It is a long-term partnership.' Vendor's experience of the RFID technology and follow-up support is essential to the success of the project.
Organisational motivation	Motivate each level of the employees to commit to the adoption of the RFID-based system by presenting them with the emerging needs for improvement and potential benefits of the RFID technology. A factory supervisor mentioned that: 'Managers should motivate the employees to perform better, helping employees adapt to and use RFID technology to increase their productivity.'
Cost/benefits evaluation	Management needs to evaluate and justify the cost and benefits of the adoption of RFID technology and how the change can align with the company's IS strategic direction. A project manager said: 'The management considers cost/ benefit evaluation is important and treats cost/benefit analysis as a long-term evaluation strategy.' An assistant project manager said: 'Intangible benefits, such as creating a culture of continuous improvement, which can indirectly enhance the business value.'
Top management support	Top management support is required to provide better access to information and resources and enhance user support, which leads to smoother implementation processes. A project manager said: 'Senior managers' support is essential for this project in order to get moral support as well as the financial and other resources support for the implementation of the project.'
User involvement	Involve the users in the design and development process so as to understand their expectations and attitude towards the new technology in a more thorough sense. A factory manager said: 'User involvement is crucial in the design and development process as the developed system will be used by the potential or current users of the system'.

(*Continued*)

Table 4.3 Continued

Extent of progress supervision	Organisation needs to keep close supervision and investigate how well the RFID technology interacts with existing organisational settings. A factory supervisor said: 'To assess the nature and extent of using the new system, a closer monitoring of existing organisational settings in which workers and system interacts is necessary.'
Staff competence and training	Choice of the RFID technology should be compatible with the staff capability and regular training programs should be provided so that the potential users would not have problems adapting to the new system. The project manager (vendor) said: 'Conducting training programs to enhance staff's capability and job competence in using RFID technology. Most of the people in the factory do not know about the new technology.'
Policy, structure and operation process compatibility	Changes in organisational policies, structures or even operation process could bring higher business value to the organisation. The production manager said: 'Policy changes including payroll policy are critical as the new RFID system can capture the real-time production information, so sewers are not able to "cheat" under the new system. In order to maintain a stable change, payment policy needs to adjust (i.e. piece rate) so that most of the sewers would not suffer at the introduction of the RFID system in the factory.'

important factors in vendor selection and adoption success. Selecting the appropriate vendor can help a company to design and develop a more customised system according to its specific characteristics and operating environment, build a trusting relationship, which leads to a smoother implementation process, and create a culture of open discussion and continuous improvement.

4.7.2 Organisational motivation

Organisational needs and motivations for performance improvement are one of the success factors identified. It is important that an organisation motivates each level of employees to commit themselves to the adoption of the RFID-based system by presenting them with the need for improvement and the potential benefits of this technology. These factors can help to raise the level of employee support for and commitment to the organisation's adoption decision. In this case, the company explained to the employees in detail the potential benefits (i.e. the increase in sewers' pay and production line visibility for better management by the production manager) that the system could bring to different positions, roles and the organisation as a whole.

4.7.3 Cost/benefit evaluation

Management needs to evaluate and justify the costs and benefits of the adoption of RFID technology and determine how the change can be aligned with the company's strategic IS direction. Cost effectiveness should be a key issue in judging the business value of this technology. It is important to note that this cost/benefit evaluation is not of financial factors alone. As indicated in the case study, the adoption of RFID technology brings a company many intangible benefits, such as creating a culture of continuous improvement, which can indirectly enhance business value. Therefore, to gain the support of top management and staff and justify the adoption decision, it is crucial that a company perform such an evaluation after the trial/pilot study. In some cases, it may need a third-party consultancy to perform this evaluation.

4.7.4 Top management support

Similar to any other new technology adoption project, the support and commitment of top management and users are influential in the adoption decision and the success of the RFID technology implementation. The support of top management provides better access to information and resources and enhances user support, thus leading to a smoother implementation process.

4.7.5 User involvement

Whereas top management support has a more strategic impact on RFID technology implementation, user support in accepting and using the system actualises the potential business value of the technology. Therefore, it is crucial to involve users in the design and development process so as to better understand their expectations for and attitude towards the RFID system. This can help to develop a system that is more user-friendly, thus promoting a higher level of user acceptance.

4.7.6 Extent of progress supervision

The close supervision of the implementation process is another success factor. RFID technology adoption is not a one-off installation project. An organisation needs to maintain close supervision and investigate how well the new technology interacts with existing organisational settings. It may want to create a special department to keep track of the system implementation process and perform regular evaluations. In the case in this study, the company held regular operational review meetings and worked with the system vendor to improve the system continuously. This ensured that the system was more compatible with the organisational structure, policies, existing IS and operating processes. Without regular reviews, the new system would be unable to bring about so many advantages. It is also important that an organisation encourages a culture of open discussion and continuous improvement. Such a culture can help to enhance the process of operational improvements that the new technology can bring.

4.7.7 Staff competence and training

The choice of an RFID system should be compatible with staff capabilities to ensure that potential users have no problems in adapting it. In this case, the major potential users were sewers, who are not well-educated. As RFID technology is easy to use, it was a good choice. Nevertheless, it is important to provide training to users to familiarise them with the use of the new technology and the functionality of the system. This can help to enhance their acceptance of the new system and boost their confidence in using it. This, in turn, may lead to greater utilisation and assimilation of the new technology and system.

4.7.8 Policy, structure and operating process compatibility

Lastly, it is essential that each aspect of the organisation, such as organisational policy and structure, be matched with the implementation of the RFID technology. The adoption of such technology is an interactive process. Of course, it is important that the technology be customised and fitted to the organisation, but it is often the case that changes in organisational policies, structures or even operating processes can create greater business value. In this case study, RFID technology adoption came along with changes in organisational structure and payroll policies. A new department was set up to monitor the implementation process of the new system, and changes in payroll policies were put in place to make use of the real-time data collected. These changes constituted the best use of the system and enhanced the benefits it brought to the organisation.

4.8 Lessons learned from the case study

This case study has provided a detailed analysis of the implementation process of an RFID-based manufacturing process management system in a garment manufacturing company. Although the main focus of the study is a case analysis of the factors that affected the success of system implementation, as a side benefit we also gained insights into the entire implementation process. Several lessons have been learned from the successful implementation of this case study. These LL can be classified by their strategic, managerial and operational aspects. They are summarised in Table 4.4 with further explanations provided in the paragraphs that follow.

Table 4.4 List of LL from the RFID-based system implementation

Strategic level	
LL1 – Choose the right partner	Choose the right project partner based on its actual experiences and capabilities in RFID implementation.
LL2 – Justify the benefits	Both hard and soft benefits should be quantified or explained and justification for them provided.
Managerial level	
LL3 – Clear objectives and expectations	An organisation should not have unrealistic requirements for a new system, but have a good understanding of RFID technology functionalities and investigate how these functionalities could enhance organisational performance.
LL4 – Obtain users' acceptance	The creation of business value from new technology depends very much on users' acceptance. If the attitudes of an organisation's members are poor towards attempts to introduce RFID technology, then introducing this technology and changing the business process will be more difficult.
LL5 – Adopt appropriate policy and structural changes	Securing the compatibility of RFID technology with an organisation's existing practices, infrastructure and policies is important in making the best use of the real-time data collected from the RFID system.
LL6 – Obtain top management support	Securing the support of top management will lead to better access to information and resources and obtain a higher level of user support.
Operational level	
LL7 – Conduct periodic operational review meetings	Periodic reviews keep track of improvements in or obstacles to the production process and encourage continuous improvement.
LL8 – Start with a trial run	Starting with a small-scale test run of the usability and functionality of the new system will help to gain initial user acceptance.
LL9 – Keep the operation simple and provide adequate training to users	The new system's design and delivery must consider the capability of potential users. Training is needed to equip these potential users with the knowledge necessary to manage the new system.
LL10 – Errors may arise from hardware failures, human error, inappropriate procedures or a poor level of support	Investigate the root causes of errors, so as to solve them and prevent them from occurring again in future.
LL11 – Systems' inter-operability	The inter-operability of the different information systems used in an organisation is important to making the best use of the real-time information collected from an RFID system.

4.8.1 Strategic level

LL1 – Choose the right partner

A key LL is the importance of choosing the right project partner based on its actual experiences and capabilities in RFID implementation (Ngai *et al.*, 2007). As RFID technology can be used to redesign business processes, the vender should have good domain knowledge and a thorough understanding of these processes in the industry. Also, a trusting relationship between the vendor and the adopter is essential, as the process of implementing new technology may require the latter to unveil certain valuable company information to the former. In this case study, the vendor was looking for a long-term relationship rather than a one-off system installation contract. Although the price of the vendor chosen was higher than that of the others, this company was willing to provide long-term support and establish a trusting relationship with the client. The vendor not only supplied hardware and software, but also a consultation service. Such a long-term relationship helps to improve and modify the process and system constantly, thus obtaining greater business value from them.

LL2 – Identify the benefits

The benefits of implementing an RFID system should be identified at the start, with both hard and soft benefits quantified if possible. Hard benefits (which can be ascribed a dollar value) are usually generated by cost reductions, process improvements and profit enhancement, whereas soft benefits (e.g. an increase in service quality) may be difficult to quantify, but should be listed and justified anyway.

4.8.2 Managerial level

LL3 – Clear objectives and expectations

Appropriate objectives and expectations for new technology are important to the success of its implementation. An organisation should not have unrealistic requirements for a new system, but should have a good understanding of the technology's functionalities and investigate how these functionalities can enhance organisational performance. Unrealistic expectations can lead to disappointment with system functionality and affect users' level of commitment towards the change. In this case, the company needed to be clear that the adoption of an RFID-based manufacturing process management system would help to uncover problems and present to managers only the real-time status of and situation on the production line. It is also important to set up a culture that encourages open discussion and continuous improvement, so as to realise the benefits of the new system.

LL4 – Obtain users' acceptance

Users' support is important. Although the adoption of RFID technology depends on the efforts of the vendor and the support of top management, the creation of business value from the new technology depends very much on users' acceptance. If the attitude of organisational members is poor with regard to an attempt to introduce this technology, then its introduction and changes in the business process will be more difficult. In this case study, the adoption of the technology came with changes in the manufacturing process that affected the system's users. These users needed to be convinced about the benefits of the RFID system, and their acceptance of the changes and feedback on them had to be obtained. This was achieved by providing them with a thorough explanation of the new technology and its benefits to potential users and proving to them through a trial run that the changes would lead to better pay cheques and efficiency.

LL5 – Adopt appropriate policy and structural changes

The compatibility of a new technology with an organisation's existing practices, infrastructure and policies is important to the success of its implementation. It is important to adopt certain changes in organisational policies or structures so that the new technology can confer greater benefits on the organisation and is not constrained by existing policies or structures. In this case, an independent support team was set up, not only to provide system maintenance but also to help identify any problems associated with implementation and raise those problems during the production review meeting.

LL6 – Obtain top management support

As in the adoption of any new technology, the support of top management is key to success. Securing that support can lead to better access to information and resources and obtain a higher level of user support. In this case, whenever problems occurred after the implementation of the new technology, the support team investigated the root causes and determined whether they were caused by system error, human error or business process and structural incompatibility. This investigation required the support of top management so that the team had better access to information and cooperation from the lower level. After the root causes have been determined, the support of top management is needed to effect the changes necessary to prevent the same errors from recurring in future.

4.8.3 Operational level

LL7 – Conduct periodic operational review meetings

It is important that management conduct periodic operational review meetings to keep track of improvement in or obstacles to the production process. Operational data and information must be provided in these meetings to determine whether further action is required. In this case, defect and lost-time analysis was reviewed regularly in the operational meetings. After the RFID system had been in place for a certain amount of time, we found that the waiting time for some of the sewing lines was as high as one-third of their total operating time. They were either waiting for cut pieces, or their machines were out of order. These problems were reported in the operational review meetings and gradually resolved.

LL8 – Start with a trial run

Because such technology affects production directly, its implementation requires care and needs to begin with a trial run. In this case, some of the sewing lines (8 lines) were installed with the RFID-based manufacturing process management system in the trial-run stage. Barcoding (the system originally used) and RFID technology ran in parallel. The data generated from the barcoding system were used to verify the data produced by the new system. This parallel run of two technologies allowed us to test the compatibility and usability of the new system. At the same time, it allowed the sewers and production line manager to adapt to the new system and payroll policies.

LL9 – Keep the operation simple and provide adequate training to users

In the garment manufacturing industry, the majority of workers (sewers) do not have a high level of education and would find it difficult to learn a new technology that was complex and technical. Therefore, it is important that operating procedures are kept simple and easy to learn. Training needed to be conducted repeatedly at the company under study, and we video-recorded the entire operation to be used as training material.

LL10 – Errors may arise from hardware failure, human error, inappropriate procedures or a poor level of support

Errors have different sources. In this case, it was common that errors arose from human mistakes, for example, made by a newcomer who was not familiar with the manufacturing process. However, it is always important to investigate the root causes of errors, so as to improve the manufacturing process and system effectiveness continuously. This can prevent the same error from recurring in the future and enhance the business value of the RFID system to the organisation.

LL11 – Systems' inter-operability

As the system vendor only provided the system with production functions, we had to design interfaces that would link it with human resources (HR) and the payroll system and the enterprise resource planning (ERP) system. The inter-operability of the different IS used in an organisation is important to make the best use of the real-time information collected from an RFID system. Data inter-changeability is another issue. Any changes in one system's database trigger changes in other systems too. It is important that the adopter involves the vendors of the different IS in coming up with a plan for data interchangeability.

An RFID-based production system able to capture real-time production and worker performance information was implemented in this case study, but more answers and further investigation are needed, and the existing business processes need further re-evaluation to make the system more effective.

4.9 Conclusion

It is likely that RFID technology will increasingly be used in both the business and manufacturing sectors. Increasing numbers of companies are trying out this technology. This study can thus provide a reference framework to help companies implement RFID technology in their own business operations. It outlines the major factors involved in the adoption intention, as well as the success factors, business value and the LL. Research and development (R&D) practitioners can learn from our experiences while carrying out the case study. This will help them to justify the benefits of RFID technology in their own organisations, strengthen their motivation to adopt such technology and prompt them to pay attention to success factors and critical issues during the development and implementation processes.

We encourage more researchers to conduct RFID-based manufacturing process management system studies to examine the success factors in different manufacturing sectors. If the factors reported herein are supported by future research, then this would suggest that they are applicable across different sectors. The identification of common success factors in the implementation of an RFID system across sectors would be an important step towards the development of a theory for the implementation of RFID innovations in manufacturing companies.

This chapter describes the R&D of an RFID-based manufacturing process management system with a focus on adoption motivation, success factors and the LL from the implementation of that system in a garment manufacturing company. Such a system can help a company to enhance production line visibility, efficiency and effectiveness, thus enhancing the business value conferred upon the organisation by the system.

We have investigated the intention to implement an RFID system through the theoretical lens of the push and pull perspective and provided a case analysis of the implementation of such a system in a garment manufacturing company that was one of the early adopters of a full RFID system in this industry. We find that both TP and NP factors influenced the intention of this company to adopt RFID technology in its factory. The former include relative advantage, compatibility, complexity, extendibility and the cost of the technology, whereas the latter include competitor and customer pressure. Eight factors that are influential to the success of RFID technology adoption in a garment manufacturing company are identified as:

1. vendor selection;
2. organisational motivation;
3. cost/benefit evaluation;
4. top management support;
5. user involvement;
6. the extent of progress supervision;
7. staff competence and training; and
8. policy, structure and operating process compatibility.

Based on our literature search, no study has yet been published in an academic journal regarding an investigation into the success factors of RFID technology adoption in the garment manufacturing industry. We believe that our research will pave the way for the adoption of the TP and NP perspective in investigating the underlying motivations behind the garment industry's investment in RFID technology adoption.

We believe that this case study will be beneficial to practitioners interested in implementing an RFID system in a manufacturing company. However, as we have provided only one case study, the generalisability of the results of the case analysis to other contexts is limited. For example, although cost is not a concern for the study subject, it might be a concern for other manufacturing companies with fewer resources. Therefore, we suggest that more in-depth case studies be conducted to analyse the differences in implementing the RFID system in different industries and that empirical studies at both the macro- and micro-levels be carried out to investigate the factors that drive RFID innovations. Future studies should also be extended to an evaluation of the business value realised by adopting RFID technology.

4.10 Acknowledgements

The authors are grateful to two anonymous referees and guest editor for their constructive comments on earlier versions of this chapter. An earlier version of this chapter was presented at the Pacific Asia Conference on Information Systems 2009. This work was supported in part by The Hong Kong Polytechnic University under grant number G-U401.

4.11 References

Akkermans, H. and Van Helden, K. (2002), Vicious and virtuous cycles in ERP implementation: A case study of interrelations between critical success factors, *European Journal of Information Systems*, 11, 35–46.

Attaran, M. (2007), RFID: An enabler of supply chain operations, *Supply Chain Management: An International Journal*, 12(4), 249–57.

Brown, I. and Russell, J. (2007), Radio frequency identification technology: An exploratory study on adoption in the South African retail sector, *International Journal of Information Management*, 27, 250–65.

Chau, P. Y. K. and Tam, K. Y. (2000), Organizational adoption of open systems: A 'technology-push, need-pull' perspective, *Information & Management*, 37, 229–39.

Chin, S., Yoon, S., Choi, C. and Cho, C. (2008), RFID+4D CAD for progress management of structural steel works in high-rise buildings, *Journal of Computing in Civil Engineering*, 22(2), 74–89.

Delen, D., Hardgrave, B. C. and Sharda, R. (2007), RFID for better supply-chain management through enhanced information visibility, *Production and Operations Management*, 16(5), 613–24.

Denzin, N. K. (1978), *The Research Act: A Theoretical Introduction to Sociological Methods*, New York: McGraw-Hill.

Green, G., Kennedy, P. and McGown, A. (2002), Management of multi-method engineering design research: A case study, *Journal of Engineering Technology and Management*, 19, 131–40.

Janz, B. D., Pitts, M. G. and Otondo, R. F. (2005), Information systems and health care. Part II: Back to the future with RFID: Lessons learnt – some old, some new, *Communications of the Association for Information Systems*, 15, 132–48.

Jones, P., Clarke-Hill, C., Hillier, D. and Comfort, D. (2005), The benefits, challenges and impacts of radio frequency identification technology (RFID) for retailers in the UK, *Marketing Intelligence & Planning*, 23(4), 395–402.

Kohn, L. T. (1997), *Methods in Case Study Analysis*, Washington, DC: The Center for Studying Health System Change.

Konomi, S. and Roussos, G. (2007), Ubiquitous computing in the real world: Lessons learnt from large-scale RFID deployments, *Personal and Ubiquitous Computing*, 11, 507–21.

Lai, K., Wong, C. W. Y. and Cheng, T. C. E. (2006), Institutional isomorphism and the adoption of information technology for supply chain management, *Computers in Industry*, 57, 93–8.

Leimeister, S., Leimeister, J. M., Knebel, U. and Krcmar, H. (2009), A cross-national comparison of perceived strategic importance of RFID for CIOs in Germany and Italy, *International Journal of Information Management*, 29, 37–47.

Lengnick-Hall, C. A., Lengnick-Hall, M. L. and Abdinnour-Helm, S. (2004), The role of social and intellectual capital in achieving competitive advantage through enterprise resource planning systems, *Journal of Engineering Technology and Management*, 21, 307–30.

Lichtenthaler, E. (2004), Technological change and the technology intelligence process: A case study, *Journal of Engineering Technology and Management*, 21, 331–48.

Little, D. and Arnst, C. (2002), Tiny radio ID tags for just pennies, *Business Week*, 18 March, 93.

Loebbecke, C., Palmer, J. and Huyskens, C. (2006), RFID's potential in the fashion industry: A case analysis. In: *Proceedings of the 19th Bled eConference*, Slovenia.

Melville, N., Kraemer, K. and Gurbaxani, V. (2004), Review – Information technology and organizational performance: An integrated model of IT business value. *MIS Quarterly*, 28(2), 283–322.

Moon, K. L. and Ngai, E. W. T. (2008), The adoption of RFID in fashion retailing: A business value-added framework, *Industrial Management & Data Systems*, 108(5), 596–612.

Ngai, E. W. T., Cheng, T. C. E., Lai, K.-H., Chai, P. Y. F., Choi, Y. S. and Sin, R. K. Y. (2007), Development of an RFID-based traceability system: Experiences and lessons learned from an aircraft engineering company, *Production and Operations Management*, 16(5), 554–68.

Ngai, E. W. T., Moon, K. K. L., Riggins, F. J. and Yi, Y. C. (2008a), RFID research: An academic literature review (1995–2005) and future research directions, *International Journal of Production Economics*, 112(2), 510–20.

Ngai, E. W. T., Suk, F. F. C. and Lo, S. Y. Y. (2008b), Development of an RFID-based sushi management system: The case of a conveyor-belt sushi restaurant, *International Journal of Production Economics*, 112, 630–45.

Ngai, E. W. T., To, C. K. M., Moon, L. K., Chan, L. K., Yeung, P. K. W. and Lee, M. C. M. (2010), RFID systems implementation: A comprehensive framework and a case study, *International Journal of Production Research*, 48(9), 2583–612.

Nystrom, P. C. (2002), Ramamurthy, K. and Wilson, A. (2002), Organizational context, climate and innovativeness: Adoption of imaging technology, *Journal of Engineering and Technology Management*, 19, 221–47.

Prasad, S. (2007) India: Will RFID tags click? *Business Week*, 31 May.

Sambamurthy, V. and Zmud, R. W. (1999), Arrangements for information technology governance: A theory of multiple contingencies, *MIS Quarterly*, 23(2), 261–90.

Sellitto, C., Burgess, S. and Hawking, P. (2007), Information quality attributes associated with RFID-derived benefits in the retail supply chain, *International Journal of Retail & Distribution Management*, 35(1), 69–87.

Sharma, A. and Citurs, A. (2005), Radio frequency identification (RFID) adoption drivers: A radical innovation adoption perspective. In: *Proceedings of the Eleventh American Conference on Information Systems*, Omaha, 1213–18.

Shepard, S. (2005), *RFID: Radio Frequency Identification*, New York: McGraw-Hill.

Shih, D. H., Chiu, Y. W., Chang, S. I. and Yen, D. C. (2008), An empirical study of factors affecting RFID's adoption in Taiwan, *Journal of Global Information Management*, 16(2), 58–80.

Stake, R. E. (1995), *The Art of Case Study Research*, Thousand Oaks, CA: Sage Publications, Inc.

Subramanian, S. K. (1990), Managing technology – The Japanese approach, *Journal of Engineering and Technology Management*, 6, 221–36.

Teo, T., Chan, C. and Parker, C. (2004), Factors affecting e-commerce adoption by SMEs: A meta-analysis. In: *Proceedings of the 2004 Australasian Conference on Information Systems*, Hobart.

Yin, R. K. (1994), *Case Study Research Design and Methods*, Thousand Oaks, CA: Sage Publications Inc.

5

The role of radio frequency identification (RFID) technologies in improving garment assembly line operations

Z. X. GUO, Sichuan University, China and W. K. WONG, S. Y. S. LEUNG, J. T. FAN and S. F. CHAN, The Hong Kong Polytechnic University, Hong Kong

DOI: 10.1533/9780857098115.99

Abstract: In this chapter, a production control problem on a flexible assembly line (FAL) with flexible operation assignment and variable operative efficiencies is described. A mathematical model of the production control problem is formulated by considering the time-constant learning curve to deal with the change of operative efficiency in real-life production. An intelligent production control decision support (PCDS) system is developed, composed of a radio frequency identification (RFID) technology-based data capture system and a PCDS model comprising a bi-level genetic optimization process, and a heuristic operation routing rule is developed. Experimental results demonstrated that the proposed PCDS system could implement effective production control decision-making.

Key words: production control, decision support system, flexible assembly lines (FALs), genetic algorithms (GAs), learning curves.

5.1 Introduction

Effective production control is useful and necessary to improve production and management performances and reduce the running costs of factories. A generic architecture for production control decision-making is shown in Fig. 5.1. In a real-life production environment, production data on production orders, production quantities of each workstation and the whole production line, operative efficiency, etc., are collected from shop floors or assembly lines by using various types of data capture methods, including the manual recording method, barcode scanning, and the most updated radio frequency identification (RFID) technology. Based on the collected production data, the production manager makes decisions to achieve various production objectives.

On shop floors or assembly lines with a low level of automation, it is impossible to obtain real-time production data owing to the absence of an effective data capture system. Thus, it is also impossible to make accurate and real-time decisions for production control. This chapter presents an intelligent production control decision support (PCDS) system, which is integrated with an RFID-based

5.1 Generic architecture for production control decision-making.

real-time data capture system, to assist in the production control decisions on a flexible assembly line (FAL).

5.2 Key issues in developing flexible assembly lines (FALs)

To meet increasingly fierce market competition, more manufacturing enterprises seek benefits from manufacturing flexibility and effective production control. Beach *et al.* (2000) provide a comprehensive review of manufacturing flexibility. There are various types of manufacturing flexibility, such as machine flexibility and routing flexibility. Machine flexibility is measured by the number of operations that a workstation processes and the time needed to switch from one operation to another. The more operations a workstation processes, the less time switching takes and the higher the machine flexibility becomes. Routing flexibility is the ability of a production system to manufacture a product using several alternative routes in the system and is usually determined by the number of such potential routes.

The FAL is an increasingly attractive assembly form for small- or mid-scale production in many industries. Unlike the traditional assembly line, some FALs allow flexible operation assignment, where one operation can be assigned to multiple workstations for processing, and multiple operations can be assigned to the same workstation. When one operation is assigned to multiple workstations, the processing of this operation is shared by the assigned workstations and is taken as a shared operation. Each shared operation of a product should be routed to an appropriate workstation on a real-time basis. Obviously, the FAL with flexible operation assignment involves machine flexibility and routing flexibility. In practice, this type of FAL is normally used in apparel manufacturing.

5.2.1 Variability of operative efficiency

On a highly automated assembly line, the efficiency to process a certain task is deterministic. Yet for FALs with a low level of automation, for example, FALs highly relying on manual effort, the operative efficiency of each task is seldom constant. The variable operative efficiency leads to fluctuation of the cycle time and increases the complexity of production control. In the field of production control, some researchers assume that the task time is an independent normal variable (Gamberini *et al.*, 2006; Moodie and Young, 1965; Suresh and Sahu, 1994). Some other researchers also assume task times with different probability distributions (Arcus, 1966; Nkasu and Leung, 1995). However, the stochastic change of the task time cannot reflect the increasing efficiency of the operator caused by repetitive and cumulative operations, and the random probability distribution cannot reflect the increasing trend of operative efficiency owing to learning effects.

Thorndike (1898) and Thurstone (1919) scientifically analyzed the learning phenomenon by focusing on the human subject's behavior and concluded that the time required for executing a specific task decreased with the cumulative experience. Some years later, learning curves were presented and widely used to describe the relationship between the operative efficiency and the accumulated operating time.

Wright (1936) established the first and most common learning curve model in 1936, which indicates that a given operation is subject to a 20% productivity improvement each time the production quantity doubles, but this curve has a significant deficiency because its asymptote is zero. After Wright's curve, various learning curve models have since been developed (Badiru, 1992), such as de Jong's equation, Wiltshire's equation and the time-constant model. On the basis of an in-depth comparison of a number of models, Hackett (1983) concluded that the time constant model (Bevis, 2004; Hitchings, 1972) is the most practical model for general use, because it can fill a wide range of observed data.

However, limited work has been done to investigate the production control problem with learning effects (Mosheiov and Sidney, 2003; Wang and Xia, 2005). The learning effect on production control decision-making for FALs has not yet been considered. For FALs that are highly reliant on manual effort, it is not unusual for many newly trained operators to be assigned to run complicated operations owing to labor shortage. Thus, it is necessary and important to consider these learning effects.

5.2.2 Production control decision-making

Research on the PCDS system has received little attention, although various decision support systems have been developed for a wide range of applications (Anon, 2007; Epic Data Inc., 2007; MSC Limited, 2007). Traditionally, the decision-making for production control relies on the experience and simple interpretation of production managers and supervisors. However, human decisions tend to be subjective, late, inconsistent and even inaccurate owing to the complexity of production decision-making problems. A large number of studies have investigated production control of the basis of two types of problems:

production scheduling and assembly line balancing. Production scheduling involves mainly shop scheduling problems (Cheng *et al.*, 1999; Framinan *et al.*, 2004; Shakhlevich *et al.*, 2000) and flexible manufacturing system scheduling (Chan and Chan, 2004; Sawik, 2002). Assembly line balancing involves mainly simple (Baybars, 1986; Scholl and Becker, 2006) and generalized assembly line balancing problems (Becker and Scholl, 2006). However, no existing literature is available on the production control and decision support system for the FAL in terms of flexible operation assignment and variable operative efficiencies.

The developed methodologies for decision support of production control involve a wide range of optimization techniques such as:

- simulation-based techniques (Chan and Chan, 2004; Chong *et al.*, 2003);
- classical optimization techniques (Crauwels *et al.*, 2005; Ibraki and Nakamura, 1994); and
- intelligent optimization algorithms (Cheng *et al.*, 1999; Guo *et al.*, 2008).

Due to the NP-hard nature of most production control problems (Gutjahr and Nemhauser, 1964), intelligent algorithms with heuristic optimization capacity are widely adopted (Charalambous and Hindi, 1991; Guo *et al.*, 2008; Scholl and Becker, 2006), while the genetic algorithm (GA) is a typical representative due to its capability of global optimization. However, none of the algorithms are applicable to any other production control problems without adjustment or modification. Moreover, due to the absence of real-time and accurate production data, most of the developed methodologies cannot be used in real-life production control.

In recent years, as the application of the RFID technology has become economically feasible, some RFID-based data capture systems have been developed to obtain real-time and accurate production data and their effectiveness has also been proved by various industrial applications and practices. With the support of effective data capture technology, it is feasible, in theory and also in practice, to develop an effective PCDS system to assist in production management for the decision-making process of production control.

In this study, the production control problem on an FAL with flexible operation assignment and variable operative efficiencies is investigated. The learning curve theory is used in this system to represent the change of operative efficiency with accumulated operating time. A GA-based intelligent PCDS system integrating with an RFID-based real-time data capture system was proposed to provide effective production decisions, which could meet the desired cycle time of each production order and minimize the total idle time on the FAL.

Section 5.2 of this chapter formulates the production control problem on an FAL and describes the variable operative efficiency based on the learning curve theory. Section 5.3 describes the GA-based intelligent PCDS system to solve the addressed problem in detail. Experiments and discussions are presented to validate the effectiveness of the proposed system in Section 5.4. Finally, the study is summarized and further research is suggested in Section 5.5.

5.3 Modelling flexible assembly lines (FALs)

In this study, the FAL is composed of a number of workstations including different types of machines. Each workstation is a physical location that accommodates an operator, a machine and a buffer. Several production orders with given quantities representing different product types are executed on the FAL. Each order comprises a series of manual operations. According to the pre-determined processing sequence, operations involved in each order must be processed in their corresponding workstations. On the FAL, one operation can be assigned to multiple workstations, while one workstation can also perform multiple operations simultaneously. The objective of production control is to meet the desired time cycle of each order and minimize the total idle time of all workstations on the FAL, by assigning and routing each operation to the most appropriate workstation or operator. In this study, it is assumed that the operation of a product cannot be interrupted once it starts. The efficiency of each operator between different operations is independent.

5.3.1 Notations

The following notation is utilized in developing the mathematical model of the production control problem addressed in this research:

1. Parameters:

   P_i, *i*th production order, $1 \leq i \leq p$
   O_{ij}, *j*th operation of order P_i
   M_{kl}, *l*th machine (workstation) of the *k*th machine type
   ST_{ij}, standard time of operation O_{ij}, the time to complete operation O_{ij} of one product with 100% operative efficiency
   η_{ijkl}, task proportion (weight) of operation O_{ij} being performed on machine M_{kl}, $0 \leq \eta_{ijkl} \leq 1$
   SO_{kl}, a set of operations, which can be processed on machine M_{kl}
   EM_{ijkl}, operative efficiency of operation O_{ij} on machine M_{kl}, which is a variable in the real-life production process
   DCT_i, desired cycle time of order P_i, the desired time interval of consecutive products entering the assembly line
   ACT_i, actual cycle time of order P_i, the actual time interval of consecutive products entering the assembly line
   α_i, penalty weight for order P_i when its actual cycle time is less than its desired cycle time
   β_i, penalty weight for order P_i when its actual cycle time is greater than its desired cycle time
   AM_i, a set of workstations processing order P_i
   N_i, the number of workstations processing order P_i
   MAT_{kl}, average assembly time of each product on machine M_{kl}
   SM_{ij}, a set of machines which can handle operation O_{ij}

C_{ij}, completion time of operation O_{ij}
ET_{ij}, elapsed time between operation O_{ij} and its latter operation including the transportation time and the set-up time
$S_{i'j'}$, starting time of operation $O_{i'j'}$
$PR(O_{i'j'}$, a set of the preceding operations of operation $O_{i'j'}$
T_{ij}, time for processing operation O_{ij}.

2. Variables:

X_{ijkl}, binary variable, X_{ijkl} is equal to 1 if operation O_{ij} is assigned to machine M_{kl}, otherwise it is equal to 0.
λ_i, binary variable, λ_i is equal to 1 if actual cycle time ACT_i is less than the desired cycle time DCT_i, otherwise it is equal to 0.

5.3.2 Mathematical model

The addressed problem minimizes two objectives and its mathematical model is described as follows:

Minimize:

$$\text{Objective 1: } Z(X_{ijkl}) = \sum_{i=1}^{p}[\alpha_i\lambda_i(DCT_i - ACT_i) + \beta_i(1-\lambda_i)(ACT_i - DCT_i)] \quad [5.1]$$

and

$$\text{Objective 2: } IT(X_{ijkl}) = \sum_{i=1}^{p}(ACT_i \cdot N_i - \sum_{kl, M_{kl}\in AM_i} MAT_{kl}) \quad [5.2]$$

where

$$MAT_{kl} = \sum_{ij, O_{ij}\in SO_{kl}} \frac{\eta_{ijkl} ST_{ij}}{EM_{ijkl}} \quad [5.3]$$

subject to

$$\sum_{kl, M_{kl}\notin SM_{ij}} X_{ijkl} = 0 \quad [5.4]$$

$$\sum_{ij} X_{ijkl} \geq 1 \quad [5.5]$$

$$\sum_{kl} X_{ijkl} \geq 1 \quad [5.6]$$

$$C_{ij} + ET_{ij} + 1 \leq S_{i'j'},\ O_{ij}\in PR(O_{i'j'}) \quad [5.7]$$

$$C_{ij} = S_{ij} + T_{ij} - 1 \quad [5.8]$$

Objective function (1) is to satisfy the desired cycle time of each order, and objective function (2) is to minimize the total idle time of all workstations on the FAL. Constraint (3) indicates that operation O_{ij} can only be operated by workstations which can handle it. Constraint (4) indicates that each workstation must process at least one operation. Constraint (5) denotes that each operation of one product must be processed. Constraint (6) indicates that each operation of one product cannot be started before its preceding operation is completed and the time-out between the two operations elapses. Constraint (7) indicates that operation O_{ij} must be assigned with the processing time.

5.3.3 Learning-curve-based operative efficiency

The operating time of operation O_{ij} in workstation M_{kl} is equal to its standard time ST_{ij} divided by the current efficiency EM_{ijkl} of the operator in workstation M_{kl}. The operative efficiency EM_{ijkl} differs between different products, owing to the increase in the accumulated operating time. In this study, the operative efficiency is represented by the time-constant learning curve model (Bevis, 2004; Hitchins, 1972), which is defined by

$$E_L(t_L) = E_b + E_\Delta(1 - e^{-tL/\tau L}) \qquad [5.9]$$

where $E_L(t_L)$ is the predicted operative efficiency at time t_L, E_b is the initial efficiency of the operator, E_Δ is the maximal improvement in performance due to learning, and τ_L is the model time constant, which is a measure of how quickly the performance improvement is achieved. The ultimate efficiency of each operator is assumed to be 100% (i.e. = 1 at $t_L \rightarrow \infty$. $E_\Delta = 1 - E_b$), and each day has 8 working hours. Figure 5.2 shows the changing trends of two learning curves with different E_b and τ_L.

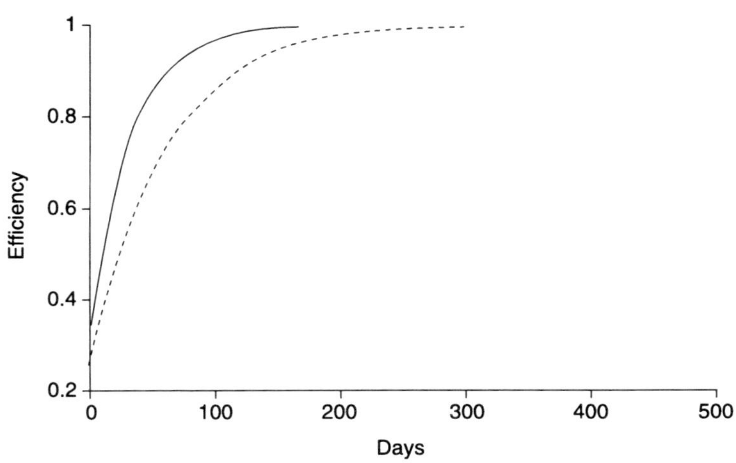

5.2 Learning curves with different E_b and τ_L: '-' = 31%, τ_L = 33 days; '---' E_b = 22%, τ_L = 58 days.

5.4 Intelligent decision support system for production control on flexible assembly lines (FALs)

An intelligent PCDS system, which can help achieve effective and real-time production control decision-making on the FAL under investigation, is introduced in this section (Fig. 5.3).

5.4.1 System architecture

As shown in Fig. 5.3, the system is composed of an RFID-based data capture system, a database server and a GA-based decision support model. The RFID-based data capture system collects all real-time job processing records and production data from the FAL. It is composed of RFID tags, RFID terminals,

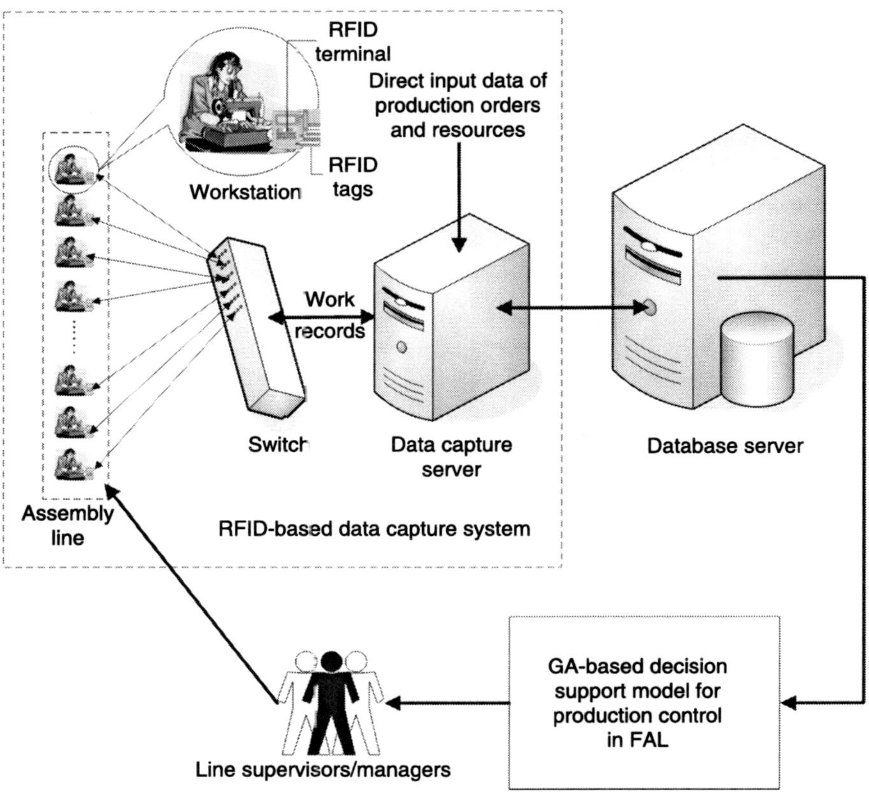

5.3 Architecture of intelligent PCDS system for FAL.

switches and data capture servers. In each workstation, an RFID terminal is installed, which can collect the job processing records by reading RFID tags attached to each batch of work-in-progress. The terminal can also display the historical job records to the operator. The terminals of each assembly line are connected to form a network by the switch. The switch is a device that channels incoming data from any of the multiple input ports to a specified output port that takes the data toward its intended destination, which is connected with a data capture server. The data communication uses the TCP/IP protocol. The data capture server collects production data based in two ways and saves them into the database.

First, the given data on production orders, workstations and assembly operators are input directly by the computer operator. Second, during the production process, each operator reads the RFID tag being attached to each batch of work-in-progress using the RFID terminal after finishing an operation. The job records from the RFID terminals are input by Ethernet. However, the data capture server also reads production information from the database and displays it on the RFID terminals. On the basis of the real-time production data stored in a database server using MySQL, Ms SQL Server or Oracle according to different requirements for data processing, the decision support model generates effective solutions to production control of the FAL. The recommended solutions are implemented by the assembly line supervisors or managers through assigning and routing operations of each product to the determined workstation on a real-time basis.

The GA-based decision support model is the kernel of the system (Fig. 5.4). As shown in the decision support model of the figure, the decision support for production control of the FAL is implemented according to the following procedures:

- *Procedure 1*: Assign flexibly operations of each type of product to different operations. One operation can be assigned to multiple workstations and multiple operations can be assigned to the same workstation. This optimized operation assignment can be implemented by a genetic optimization process with an operation-based representation.
- *Procedure 2*: Determine the optimal task proportions of each shared operation being processed in different workstations, which can be implemented by using a real-coded GA.
- *Procedure 3*: According to the optimized operation assignment results and task proportions of each shared operation, route operations of each product to the appropriate workstation on a real-time basis, which can be implemented by using a heuristic routing rule.

For each operation assignment solution in the first procedure, the second procedure seeks the optimal task proportions of each shared operation. In other words, the solution to the second procedure relies on the solution to the first procedure. It is a typical bi-level optimization problem. In this study, the bi-level GA (BiGA)

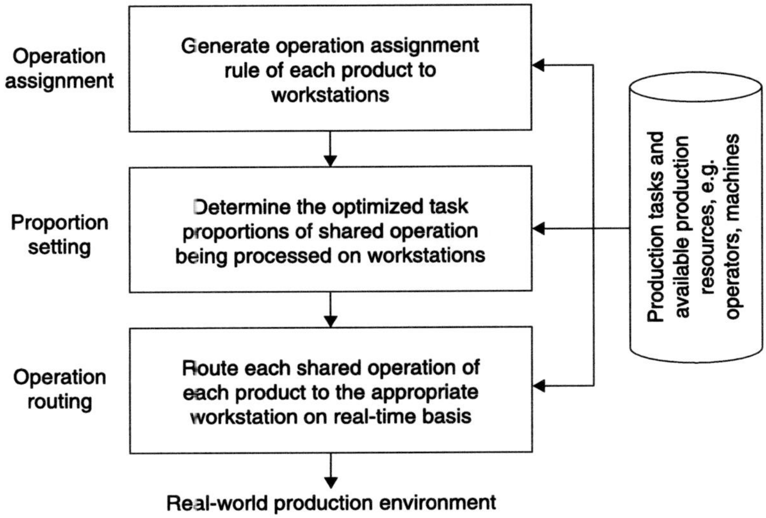

5.4 GA-based decision support model for production control on FAL.

presented by Guo *et al.* (2008) is modified to implement the two procedures. In the BiGA, a novel representation and modified genetic operators are presented to deal with the flexible operation assignment.

5.4.2 Modified bi-level genetic algorithm

The modified BiGA (mBiGA) is also composed of two genetic optimization processes (GA-1 and GA-2) (Fig. 5.5). In the mBiGA, GA-1 generates the optimal operation assignment to workstations, in which the representation and genetic operators are the same as those in GA-1 of the BiGA (Guo *et al.*, 2008). Based on each chromosome of GA-1, GA-2 determines the task proportions (weights) of the shared operation being processed in different workstations. Seeking the optimal task proportions (weights) is a first-order multivariate function optimization problem, which can be optimized by a GA with real-coded representation. The following processes of the BiGA are different from those of the mBiGA:

1. *Representation in GA-2*: each gene represents the task proportion of an operation assigned to the corresponding workstation. By considering the assignment of n operations, nm_{ij} denotes the number of machines allocated to process operation O_{ij} and PS_{ij} denotes the summation of nm_{ij}-1 weights of O_{ij}. The number of genes in each chromosome of GA-2 is the summation of nm_{ij} minus n, since the nm_{ij}^{th} weight is equal to 1-PS_{ij}.

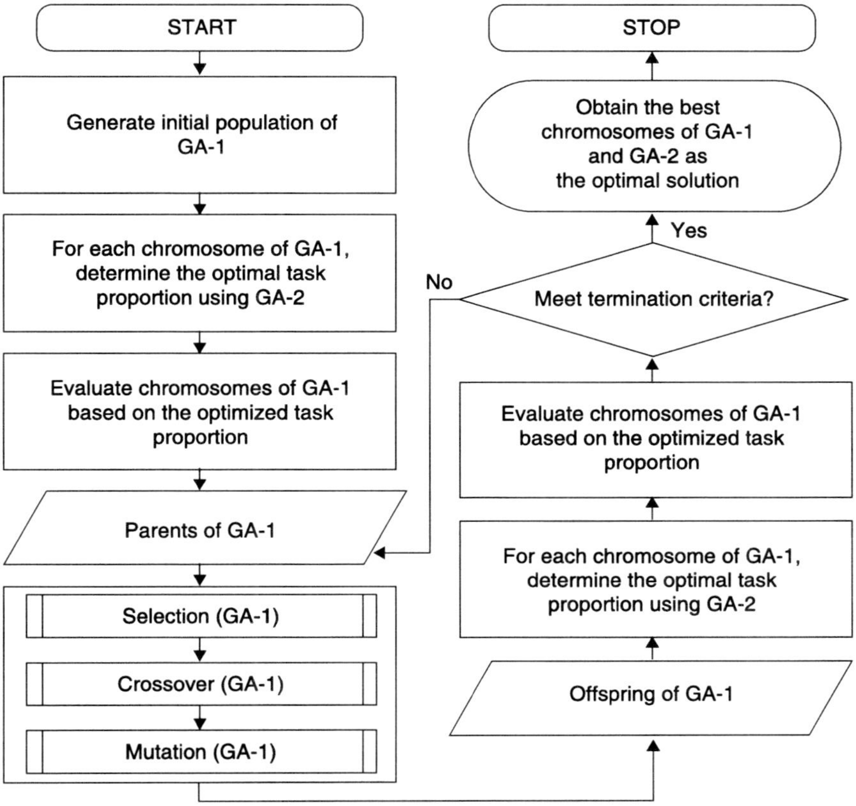

5.5 Structure of the modified bi-level genetic algorithm.

2. *Initialization in GA-2*: the initial population is generated by initializing each task proportion (weight) randomly in the chromosome between 0 and 1, based on the premise of $PS_{ij} \leq 1$.
3. *Fitness function*: the fitness function of GA-1 is the same as that of GA-2.

The objective of addressing the production control problem is to satisfy the desired cycle time of each order and minimize the total idle time of all workstations, which can be defined by

$$OBJ(X_{ijkl}) = w_z \cdot Z(X_{ijkl}) + w_{IT} \cdot IT(X_{ijkl}) \quad [5.10]$$

where w_z and w_{IT} are the relative weights placed upon the objectives $Z(X_{ijkl})$ and $IT(X_{ijkl})$, respectively. The less the weighted summation of the two objectives, the greater the fitness becomes. The fitness function *ft* can be defined as:

$$ft = \frac{100}{OBJ(X_{ijkl}) + 1} \quad [5.11]$$

Based on the chromosomes in GA-1 and GA-2, the fitness should be calculated in each generation. In this study, the operative efficiency of the operator is variable and affected by the learning phenomenon. Therefore, the operative efficiency of one operator processing the same operation differs between different production cycles. It is assumed that the order size (product quantity) of order P_i is OS_i and the number of assembly operation is U_i. The chromosomes in GA-1 and GA-2 are given. On the basis of chromosomes of GA-1 and GA-2, the procedure to calculate fitness is described in detail as follows:

1. Parameter initialization: order index $i=1$;
2. Parameter initialization: for order P_i, initialize product index $u=1$, operation index $v=1$, production days $iDays=0$;
3. For operation v of the uth product, select an operator (operator w) to process it according to the corresponding assignment rule and the operator's task proportion assigned;
4. For operator w, calculate his/her operating time for processing the operation v of the current product:
   i. Calculate his/her accumulated operating time $AccT_1$ for processing operation v;
   ii. Calculate his/her accumulated operating time $AccT_2$ on the current day;
5. If $AccT_2 > 8*3600$ (working time in second unit per day), then $iDays = iDays + 1$, $AccT_2 = 0$;
6. $v = v + 1$. If $v > U_i$, then go to (7), otherwise go to (3);
7. $u = u + 1$. If $u > OS_i$, then go to (8), otherwise set $v=1$ and go to (2);
8. On the basis of the accumulated operating time $AccT_1$ of each operation, calculate the actual cycle time of order P_i and the idle time of workstations processing order P_i;
9. $i = i + 1$. If i is greater than the number of orders, i.e. $i > p$, then go to (10), otherwise go to (1);
10. On the basis of the actual cycle time of all orders and the total idle time of all workstations, calculate the fitness.

5.4.3 Operation routing rule

After the preceding operations of the shared operation of a product are completed, the next step is to route the shared operation to an appropriate workstation based on the optimized operation assignment and task proportions of the shared operation being processed in different workstations. Assuming that operation O_{ij} is assigned to m machines ($M_{k1}, M_{k2}, \ldots M_{km}$) according to the optimized operation assignment, η'_{ijkl} denotes the optimized task proportion that operation O_{ij} should be processed on machine M_{kl} (η'_{ijkl}>0), η''_{ijkl} denotes the task proportion that operation O_{ij} has been processed on machine M_{kl} and Q_{ijkl} denotes the number of operation O_{ij}, which has been assigned to machine M_{kl}. For shared operation O_{ij} of a product, the heuristic operation routing rule is described in the following procedure:

1. Calculate $\eta''_{ijkl} = Q_{ijkl} / (\sum_{l=1}^{n} Q_{ijkl})$ for each machine M_{kl} (for the first product, set $\eta''_{ijkl}=0$).
2. Calculate $\eta''_{ijkl}/\eta'_{ijkl}$ for each machine M_{kl}.
3. Assign operation O_{ij} of the current product to the machine M_{kl} with the minimum $\eta''_{ijkl}/\eta'_{ijkl}$. If multiple machines have the same minimum value, one of the machines will be chosen randomly.

Table 5.1 shows an example of the operation routing to process operation O_{11} of 10 units of the same product. Operation O_{11} is assigned to machines M_{11}, M_{12} and M_{13}. The task proportions of operation O_{11} to be processed on these three machines are 0.3, 0.3 and 0.4 respectively, generated by the proposed mBiGA. The rows of $\eta''_{ijkl}/\eta'_{ijkl}$ describe the current value $\eta''_{ijkl}/\eta'_{ijkl}$ of operation O_{11} of each product in the relevant machine, and the shaded grid represents that the corresponding machine is selected to process the operation of the corresponding product. According to the results of operation routing (Table 5.1), operation O_{11} of the first unit of the product is assigned to M_{11}, that of the second unit of the product is assigned to M_{13}, etc. After the 10 units of the product are completed, the actual task proportion processed on each machine is equal to the optimized task proportion.

Table 5.1 Example of operation routing to process operation O_{11} of 10 products

Product No.		1	2	3	4	5	6	7	8	9	10
$\eta''_{ilkj}/\eta'_{ilkj}$	M_{11}	0	0	1.667	1.1	0.833	1.333	1.111	0.952	1.25	1.111
	M_{12}	0	0	0	1.1	0.833	0.667	1.111	0.952	0.833	1.111
	M_{13}	0	2.5	1.25	0.825	1.25	1	0.833	1.071	0.938	0.833

5.5 Testing the effectiveness of the intelligent production control decision support (PCDS) system

To validate the effectiveness of the proposed intelligent PCDS system for production control of the addressed FAL, a series of experiments were conducted. This section presents the details of these experiments and examines the effect of the learning phenomenon on production control decision-making. In these experiments, assume there is no shortage of materials, workstation breakdown or operator absenteeism on the FAL. The FAL discussed is empty initially. In other words, there is no work-in-process in each workstation.

5.5.1 Experiments

Under normal conditions, the operative efficiency of each operator increases with the increase in the accumulated operating time on the FAL due to the accumulated learning effect. To investigate the influence of efficiency increase on production decisions, two orders of different product quantities based on two different cases were executed in each experiment. In Case 1, 1000 products were made, while 5000 products were made in Case 2. In each experiment, the two orders were scheduled for production and the basic data for the orders are listed as follows:

1. *Experiment 1*: the desired cycle time of both orders 1 and 2 was 400 s. The product's assembly process of Order 1 was from operations 1 to 7, and Order 2 was from operations 8 to 12.
2. *Experiment 2*: the desired cycle time of Orders 1 and 2 was 55 and 130 s, respectively. The product's assembly process of Order 1 was from Operations 1 to 6, and Order 2 was from Operations 7 to 11.
3. *Experiment 3*: the desired cycle time of both orders was 50 s. The assembly operations of the two orders were the same as those in Experiment 2.
4. *Experiment 4*: the desired cycle time of Orders 1 and 2 was 70 and 225 s, respectively. The product's assembly process of Order 1 was from Operations 1 to 5, and Order 2 was from Operations 6 to 10.

The experiments were conducted on an FAL with 11 workstations, Workstations 1 to 8 used machines of type 1 and workstations 9 to 11 used machines of type 2. The operative efficiency of each workstation depended on the type of machine, the skill level and the recent performance of the operator. The operative efficiencies in the four experiments are shown in Tables 5.2 to 5.5. The efficiency is set at 0 if the operator cannot process the corresponding operation. The standard time of each operation in the experiments is shown in the last rows of Tables 5.2 to 5.5. The processing time of operation O_{ij} in workstation M_{kl} is equal to the standard time of this operation divided by its operative efficiency in workstation M_{kl}.

In this study, the learning curves of different operators were probably different and each operator had only one learning curve for different operations. That is, whichever operation was processed, the learning curve of the operator was the same. The parameters of the learning curves of each operator are shown in Table 5.6.

The production control solutions to the four experiments generated by the proposed PCDS system are shown in Tables 5.7 and 5.8. In Table 5.7, the first row represents the machine type, the second shows the workstation number, and other rows show the optimized operation assignment and task proportions of different experiments to the workstation. The first value of each cell represents the operation

Table 5.2 Basic data of experiment 1

Machine type	Workstation No.	Operation No. of order 1							Operation No. of order 2				
		1	2	3	4	5	6	7	8	9	10	11	12
Type 1	1	90%	100%	95%	90%	0	0	100%	90%	0	100%	0	95%
	2	70%	75%	70%	70%	0	0	75%	80%	0	80%	0	70%
	3	80%	80%	80%	85%	0	0	80%	80%	0	75%	0	85%
	4	70%	70%	65%	60%	0	0	70%	70%	0	75%	0	65%
	5	80%	85%	75%	80%	0	0	85%	90%	0	95%	0	80%
	6	90%	90%	85%	90%	0	0	90%	90%	0	90%	0	95%
	7	95%	95%	100%	100%	0	0	95%	90%	0	85%	0	100%
	8	80%	75%	75%	75%	0	0	80%	85%	0	75%	0	80%
Type 2	9	0	0	0	0	100%	95%	0	0	100%	0	95%	0
	10	0	0	0	0	70%	80%	0	0	85%	0	75%	0
	11	0	0	0	0	90%	90%	0	0	90%	0	85%	0
Standard time (s/piece)		308	310	335	315	320	302	280	310	180	320	125	325

Table 5.3 Basic data of experiment 2

Machine type	Workstation No.	Operation No. of order 1						Operation No. of order 2				
		1	2	3	4	5	6	7	8	9	10	11
Type 1	1	90%	100%	95%	0	90%	100%	90%	0	100%	0	95%
	2	70%	75%	70%	0	70%	75%	80%	0	80%	0	70%
	3	80%	80%	80%	0	85%	80%	80%	0	75%	0	85%
	4	70%	70%	65%	0	60%	70%	70%	0	75%	0	65%
	5	80%	85%	75%	0	80%	85%	90%	0	95%	0	80%
	6	90%	90%	85%	0	90%	90%	90%	0	90%	0	95%
	7	95%	95%	100%	0	100%	95%	90%	0	85%	0	100%
	8	80%	75%	75%	0	75%	80%	85%	0	75%	0	80%
Type 2	9	0	0	0	100%	0	0	0	100%	0	95%	0
	10	0	0	0	70%	0	0	0	85%	0	75%	0
	11	0	0	0	90%	0	0	0	90%	0	85%	0
Standard time (s/piece)		25	36	30	75	45	54	125	58	132	65	120

Table 5.4 Basic data of experiment 3

Machine type	Workstation No.	Operation No. of order 1						Operation No. of order 2				
		1	2	3	4	5	6	7	8	9	10	11
Type 1	1	90%	100%	95%	0	90%	100%	90%	0	100%	0	95%
	2	70%	75%	70%	0	70%	75%	80%	0	80%	0	70%
	3	80%	80%	80%	0	85%	80%	80%	0	75%	0	85%
	4	70%	70%	65%	0	60%	70%	70%	0	75%	0	65%
	5	80%	85%	75%	0	80%	85%	90%	0	95%	0	80%
	6	90%	90%	85%	0	90%	90%	90%	0	90%	0	95%
	7	95%	95%	100%	0	100%	95%	90%	0	85%	0	100%
	8	80%	75%	75%	0	75%	80%	85%	0	75%	0	80%
Type 2	9	0	0	0	100%	0	0	0	100%	0	95%	0
	10	0	0	0	70%	0	0	0	85%	0	75%	0
	11	0	0	0	90%	0	0	0	90%	0	85%	0
Standard time (s/piece)		25	36	30	75	45	54	38	20	35	25	40

Table 5.5 Basic data of experiment 4

Machine type	Workstation No.	Operation No. of order 1						Operation No. of order 2			
		1	2	3	4	5	6	7	8	9	10
Type 1	1	90%	95%	0	100%	90%	90%	0	100%	0	95%
	2	70%	70%	0	70%	75%	80%	0	80%	0	70%
	3	80%	80%	0	85%	80%	80%	0	75%	0	85%
	4	65%	75%	0	70%	60%	75%	0	70%	0	65%
	5	80%	75%	0	85%	80%	90%	0	95%	0	80%
	6	85%	90%	0	90%	90%	90%	0	90%	0	95%
	7	100%	95%	0	95%	100%	85%	0	90%	0	100%
	8	80%	75%	0	80%	75%	80%	0	75%	0	85%
Type 2	9	0	0	95%	0	0	0	100%	0	100%	0
	10	0	0	75%	0	0	0	85%	0	70%	0
	11	0	0	85%	0	0	0	90%	0	90%	0
Standard time (s/piece)		20	75	52	30	90	180	160	408	240	205

Table 5.6 Parameters of learning curve of each operator

Workstation No.	1	2	3	4	5	6	7	8	9	10	11
E_b	22%	31%	44%	20%	24%	26%	26%	38%	28%	26%	32%
τ_L (day)	58	33	46	42	56	52	41	26	54	41	28

number and the value in the bracket represents the task proportion η_{ijkl} of the operation being processed in the corresponding workstation. For example, the value 10(1) in the row of 'Case 1 of Experiment 1' shows that workstation 2 processes the whole (100%) of Operation 10, and the value (5(0.13),6(1)) shows that workstation 9 processes 13% of Operation 5 and 100% of Operation 6. In Table 5.8, the columns of 'Actual cycle time' show the optimized actual cycle time (seconds) of Orders 1 and 2 in four experiments, whereas the columns of 'Idle time' and 'Line efficiency' show the optimized average idle time (seconds) in each cycle and the optimized line efficiencies of Orders 1 and 2 in the four experiments. In this study, the line efficiency of order P_i is defined as the average processing time of workstations processing this order in each cycle divided by the actual cycle time of this order.

As shown in Table 5.7, the proposed mBiGA can implement flexible operation assignments, including assigning one operation to different workstations and multiple operations to the same workstation. For instance, in Case 1 of Experiment 2, Operation 7 is assigned to workstations 2 and 8, while Operations 7 and 9 are assigned to workstation 2. At the same time, the different task proportions of the shared operation are also optimized. For instance, in Case 1 of Experiment 2, the processing of Operation 7 is shared in workstations 2 and 8 and the task proportions are 0.67 and 0.33, respectively. In different cases of each experiment, the operation assignments are different because the quantities of the processed products are different.

Table 5.7 indicates that the actual cycle time of two orders are close to the desired cycle time in each case and the assembly line efficiency is also very good, which is between 87.48 and 98.12%. For instance, for Case 1 of Experiment 4, the actual cycles of Orders 1 and 2 are 69.99 and 224.98, respectively, that is their percentage errors are only 0.014 and 0.009%, respectively. For Order 1, the total idle time of all workstations in each cycle is only 6.58, which is less than 9.4% of its cycle time. The production flow of this order is thus very smooth. These results demonstrate that the proposed decision support system can solve the addressed production control problem effectively.

Table 5.8 shows that the actual cycle time of each order in Case 2 is less than that in Case 1. The operative efficiency of operators increases with the increase in the accumulated operating time. Therefore, in Case 2, more operating time is accumulated, leading to higher operative efficiency and lower cycle time.

Table 5.7 Optimized operation assignment and task proportions of four experiments (with learning effects)

Machine type			Type 1						Type 2				
Workstation No.			1	2	3	4	5	6	7	8	9	10	11
Operation Assignment	Experiment 1	Case 1	12(1)	10(1)	2(1)	7(1)	4(1)	3(1)	8(1)	1(1)	5(0.13), 6(1)	5(0.87)	9(1), 11(1)
		Case 2	3(1)	2(1)	8(1)	7(1)	12(1)	4(1)	1(1)	10(1)	5(0.07), 6(1)	5(0.93)	9(1), 11(1)
	Experiment 2	Case 1	6(1)	7(0.67), 9(0.15)	9(0.16), 11(0.49)	9(0.69)	3(0.05), 5(0.83)	2(1), 5(0.17)	1(1), 3(0.95)	7(0.33), 1 1(0.51)	8(1), 10(1)	4(0.5)	4(0.5)
		Case 2	6(1)	9(0.65)	9(0.35), 11(0.49)	7(0.64)	7(0.36), 11(0.51)	2(1), 5(0.3)	3(0.51), 5(0.7)	1(1), 3(0.49)	8(1). 10(1)	4(0.51)	4(0.49)
	Experiment 3	Case 1	7(1)	1(1), 6(0.19)	9(1)	3(1)	11(1)	5(1)	6(0.81)	2(1)	4(0.54), 10(0.36)	4(0.46)	8(1), 10(0.64)
		Case 2	7(1)	1(1), 6(0.21)	9(1)	3(1)	11(1)	5(1)	6(0.79)	2(1)	4(0.52), 10(0.41)	4(0.48)	8(1), 10(0.59)
	Experiment 4	Case 1	6(0.52), 10(0.52)	1(0.47), 5(0.47)	6(0.48), 10(0.48)	2(0.49), 4(0.49)	8(0.51)	1(0.53), 5(0.53)	8(0.49)	2(0.51), 4(0.51)	7(0.24), 9(0.66)	3(1)	7(0.76), 9(0.34)
		Case 2	6(0.55), 10(0.53)	6(0.45), 10(0.47)	1(0.51), 5(0.51)	2(0.49), 4(0.50)	8(0.52)	1(0.49), 5(9.49)	8(0.48)	2(0.51), 4(0.50)	7(0.50), 9(0.50)	3(1)	7(0.50), 9(0.50)

Table 5.8 Optimized production control results of four experiments (with learning effects)

		Actual cycle time		Idle time		Line efficiency	
		Order 1	Order 2	Order 1	Order 2	Order 1	Order 2
Experiment 1	Case 1	399.30	395.74	271.58	74.02	90.28%	95.32%
	Case 2	391.99	392.52	343.53	149.64	87.48%	90.47%
Experiment 2	Case 1	54.94	129.94	28.47	49.51	91.36%	92.38%
	Case 2	54.74	129.22	28.82	65.44	91.23%	89.87%
Experiment 3	Case 1	49.99	49.99	12.97	20.56	96.29%	91.77%
	Case 2	49.94	49.94	15.91	23.54	95.45%	90.57%
Experiment 4	Case 1	69.99	224.98	6.58	49.78	98.12%	96.31%
	Case 2	69.66	224.96	12.33	44.77	96.46%	96.68%

5.5.2 Influence of different initial operative efficiencies

In this section, experimental Cases 3 to 5 are presented for the investigation of different initial operative efficiencies. The data of the three cases are the same as those of Case 2 of Experiment 1, except for the operative efficiencies of each operator. In Cases 3 to 5, the initial operative efficiencies of each operator are equal to his/her operative efficiencies (Table 5.2) multiplied by 90, 80 and 70%, respectively. For instance, the efficiencies of the operator of workstation 1 processing Operations 1 to 3 should be 81, 90 and 85.5%, respectively in Case 3. Tables 5.9 and 5.10 show the optimized production decision and control results of the three cases. Although the initial operative efficiencies are different, the optimized production control results of Cases 3 and 4 are still very good, in which the desired cycle time is met and the line efficiencies are greater than 94%. As for Case 5, although the actual production cycle of Order 1 lags behind the desired cycle time, because the initial efficiencies are too low to reach the desired production capacity, the results of generated idle time and line efficiency are still good. These results also demonstrate the effectiveness of the proposed decision support method.

5.5.3 Experiment without the consideration of learning effects

In this section, we present the production control results of Case 2 of the four experiments without learning effects. The optimized production decision and control results are shown in Tables 5.11 and 5.12. The results are also good and the actual cycle time of Case 2 of Experiments 1 and 3 is even the same as the desired time.

Table 5.9 Optimized operation assignment and task proportions (additional cases of experiment 1)

Machine type		Type 1						Type 2				
Workstation No.		1	2	3	4	5	6	7	8	9	10	11
Operation Assignment of Experiment 1	Case 3	8(1)	3(0.49), 7(0.46)	10(1)	1(0.26), 4(0.66)	12(1)	3(0.51), 7(0.54)	2(1)	1(0.74), 4(0.34)	5(0.52), 6(0.52)	5(0.48) 6(0.48)	9(1), 11(1)
	Case 4	3(1)	12(1)	8(1)	2(1)	4(1)	1(0.5), 10(0.5)	7(1)	1(0.5), 10(0.5)	5(1), 6(0.07)	6(0.93)	9(1), 11(1)
	Case 5	8(1)	1(0.77), 4(0.18)	3(0.81), 7(0.09)	2(1)	10(1)	12(1)	1(0.23), 4(0.82)	3(0.19), 7(0.91)	5(0.85), 6(0.18)	5(0.15), 6(0.82)	9(1), 11(1)

Table 5.10 Optimized results of production control (additional cases of experiment 1)

		Actual cycle time		Idle time		Line efficiency	
		Order 1	Order 2	Order 1	Order 2	Order 1	Order 2
Experiment 1	Case 3	399.49	388.17	164.71	77.85	94.11%	94.99%
	Case 4	398.70	398.70	154.82	48.97	95.15%	97.54%
	Case 5	411.19	399.19	30.90	17.14	99.41%	99.14%

Without learning effects, the operative efficiencies of each operator were unchanged and equal to the operator's initial efficiencies before production. In other words, the maximum production capacity of production system without learning effects was less than that with learning effects. Even if the production capacity of the FAL decreased, the proposed methodology could still generate effective production control decisions.

In the above experiments, on the basis of different production tasks and production situations, different operation assignments and task proportions were generated. Whichever production tasks and production situations were considered, the optimized production decisions could meet the production objectives, which shows the effectiveness of the proposed decision support system. Moreover, based on the same production task, the generated production decisions with learning effects were different from those without learning effects, owing to the increase of operative efficiencies. Since the production decision without considering learning effect is far from production practice, these effects must be considered in both theory and practice.

The optimized results of this study were obtained based on the following settings:

- the population sizes of GA-1 and GA-2 were 200 and 100, respectively;
- the maximum numbers of generations of GA-1 and GA-2 were 100 and 50, respectively;
- the penalty weights α_i and β_i of each order were 1 and 10; and
- the relative weights w_z and w_{IT} were both set at 1.

5.6 Conclusion

This study investigates the production control problem on an FAL, so as to meet the desired cycle time of each order and minimize the total idle time of all workstations on the FAL. The mathematical model of the addressed problem was presented and time-constant learning curve model was adopted to describe the variable operative efficiencies on the FAL. An intelligent decision support system

Table 5.11 Optimized operation assignment and task proportions of Case 2 of four experiments (without learning effects)

Machine type		Type 1						Type 2				
Workstation No.		1	2	3	4	5	6	7	8	9	10	11
Operation Assignment	Experiment 1	12(1)	10(1)	2(1)	7(1)	4(1)	3(1)	8(1)	1(1)	5(0.13), 6(1)	5(0.87)	9(1), 11(1)
	Experiment 2	6(1)	7(0.67), 9(0.15)	9(0.16), 11(0.49)	9(0.69)	3(0.05), 5(0.83)	2(1), 5(0.17)	1(1), 3(0.95)	7(0.33), 11(0.51)	8(1), 10(1)	4(0.5)	4(0.5)
	Experiment 3	5(1)	1(1), 6(0.15)	9(1)	3(1)	11(1)	7(1)	6(0.85)	2(1)	4(0.64), 10(0.07)	4(0.36)	8(1), 10(0.93)
	Experiment 4	6(0.60), 10(0.45)	1(0.87), 5(0.36)	6(0.40), 10(0.55)	2(0.52), 4(0.42)	8(0.51)	1(0.13), 5(0.64)	8(0.49)	2(0.48), 4(0.58)	7(0.23), 9(0.67)	3(1)	7(0.77), 9(0.33)

Table 5.12 Optimized production control results of Case 2 of four experiments (without learning effects)

	Actual cycle time		Idle time		Line efficiency	
	Order 1	Order 2	Order 1	Order 2	Order 1	Order 2
Experiment 1	400.00	400.00	82.42	166.39	97.06%	89.60%
Experiment 2	54.82	129.44	27.65	47.52	91.59%	92.66%
Experiment 3	50.00	50.00	22.60	11.69	93.54%	95.32%
Experiment 4	70.00	224.89	5.80	46.59	98.34%	96.60%

was developed to address the production control problem, in which an RFID-based data capture system was presented to collect the real-time production data from the FAL and a PCDS model was presented to assist in production control decisions on the FAL. In the PCDS model, the mBiGA was used to generate the operation assignment to workstations and task proportions of each shared operation being processed in different workstations. A heuristic operation routing rule was also developed to route the shared operation of each product to an appropriate workstation on a real-time basis. Experimental results were presented to validate the effectiveness of the proposed decision support system. The results confirm that the learning phenomenon should be considered in production control decision-making, because it can result in the increase of operative efficiency and production performance.

This study considers the change of operative efficiency based on the learning curve theory. Since the change of operative efficiency can also be influenced by other factors such as negligence, relearning, and status of machine and operator, future research can focus on the effects of these factors on production control decision-making on the FAL and other production systems.

5.7 Acknowledgement

The authors would like to thank Genexy Company Limited for providing the industrial data and financial support in this research project (Project No. ZW90). Reprinted from *Expert Systems with Applications*, 36(3), Part 1, Z. X. Guo, W. K. Wong, S. Y. S. Leung, Fan, J. T. and S. F. Chan, Intelligent production control decision support system for flexible assembly lines, 4268-77. Copyright (2009), with permission from Elsevier.

5.8 References

Anon (2007), ERP systems can achieve up to 50% higher return on investment when used with RFID mobile data capture systems, *Assembly Automation*, 27(1), 74–5.

Arcus, A. L. (1966), COMSOAL: A computer method of sequencing operations for assembly lines, *International Journal of Production Research*, 4(4), 259–77.

Badiru, A. (1992), Computational survey of univariate and multivariate learning-curve models, *IEEE Transactions on Engineering Management*, 39(2), 176–88.

Baybars, I. (1986), A survey of exact algorithms for the simple assembly line balancing problem, *Management Science*, 32(8), 909–32.

Beach, R., Muhlemann, A. P., Price, D. H. R., Paterson, A. and Sharp, J. A. (2000), A review of manufacturing flexibility, *European Journal of Operational Research*, 122(1), 41–57.

Becker, C. and Scholl, A. (2006), A survey on problems and methods in generalized assembly line balancing, *European Journal of Operational Research*, 168(3), 694–715.

Bevis, F. W. (1970), An exploratory study of industrial learning with special reference to work study standards, MSc Thesis, University of Wales.

Chan, F. and Chan, H. (2004) A comprehensive survey and future trend of simulation study on FMS scheduling, *Journal of Intelligent Manufacturing*, 15(1), 87–102.

Charalambous, O. and Hindi, K. (1991), A review of artificial intelligence-based job-shop scheduling systems, information and decision technologies, 17(3), 189–202.

Cheng, R., Gen, M. and Tsujimura, Y. (1999), A tutorial survey of job-shop scheduling problems using genetic algorithms. Part II: Hybrid genetic search strategies, *Computers & Industrial Engineering*, 37(1–2), 51–5.

Chong, S. C., Sivakumar, A. I. and Gay, R. K. L. (2003), Simulation-based scheduling for dynamic discrete manufacturing, *Proceedings of the 2003 Winter Simulation Conference*, New Orleans, LA.

Crauwels, H., Potts, C., Van Oudheusden, D. and Van Wassenhove, L. (2005), Branch and bound algorithms for single machine scheduling with batching to minimize the number of late jobs, *Journal of Scheduling*, 8(2), 161–77.

Epic Data Inc., RFID Systems. Available from: http://www.epicdata.com/data/rfid-systems.php, 2007

Framinan, J., Gupta, J. and Leisten, R. (2004), A review and classification of heuristics for permutation flow-shop scheduling with makespan objective, *Journal of the Operational Research Society*, 55(12), 1243–55.

Gamberini, R., Grassi, A. and Rimini, B. (2006), A new multi-objective heuristic algorithm for solving the stochastic assembly line re-balancing problem, *International Journal of Production Economics*, 102(2), 226–43.

Guo, Z., Wong, W., Leung, S., Fan, J. and Chan, S. (2008), A genetic-algorithm-based optimization model for scheduling flexible assembly lines, *International Journal of Advanced Manufacturing Technology*, 36(1–2): 156–68.

Gutjahr, A. L. and Nemhauser, G. L. (1964), An algorithm for the line balancing problem, *Management Science*, 11(2), 308–15.

Hackett, E. (1983), Application of a set of learning-curve models to repetitive tasks, *Radio and Electronic Engineer*, 53(1), 25–32.

Hitchings, B. (1972), Dynamic Learning curve models describing the performance of human operators on repetitive industrial tasks, MSc Thesis, University of Wales.

Ibraki, T. and Nakamura, Y. (1994), A dynamic-programming method for single-machine scheduling, *European Journal of Operational Research*, 76(1), 72–82.

Moodie, C. L. and Young, H. E. (1965), A heuristic method of assembly line balancing for assumptions of constant or variable work element times, *Journal of Industrial Engineering*, 16(1), 23–9.

Mosheiov, G. and Sidney, J. (2003), Scheduling with general job-dependent learning curves, *European Journal of Operational Research*, 147(3), 665–70.

MSC Limited (2007), Solution of manufacturing information and management in multi variety and small batch age, *Textile & Clothing*, 19(1), 58–60.

Nkasu, M. and Leung, K. (1995), A stochastic approach to assembly-line balancing, *International Journal of Production Research*, 33(4), 975–91.

Sawik, T. (2002), Monolithic vs. hierarchical balancing and scheduling of a flexible assembly line, *European Journal of Operational Research*, 143(1), 115–24.

Scholl, A. and Becker, C. (2006), State-of-the-art exact and heuristic solution procedures for simple assembly line balancing, *European Journal of Operational Research*, 168(3), 666–93.

Shakhlevich, N., Sotskov, Y. and Werner, F. (2000), Complexity of mixed shop scheduling problems: A survey, *European Journal of Operational Research*, 120(2), 343–51.

Suresh, G. and Sahu, S. (1994), Stochastic assembly-line balancing using simulated annealing, *International Journal of Production Research*, 32(8), 1801–10.

Thorndike, E. L. (1898), Animal intelligence: An experimental study of the associative processes in animals, *The Psychological Review: Ser. Monograph Supplements*, 2(8), 1–109.

Thurstone, L. L. (1919), The learning curve equation, *Psychological Monographs*, 26(114), 51.

Wang, J. and Xia, Z. (2005), Flow-shop scheduling with a learning effect, *Journal of the Operational Research Society*, 56(11), 1325–30.

Wright, T. (1936), Factors affecting the cost of airplanes, *Journal of Aeronautical Science*, 3(4), 122–8.

6

Improving order allocation in fashion supply chains using radio frequency identification (RFID) technologies

M. TEUCKE and B. SCHOLZ-REITER,
University of Bremen, Germany

DOI: 10.1533/9780857098115.126

Abstract: Apparel retail for fashion products has to cope with many different product variants, long transport distances for ready-made products, long production and delivery times, short product sales times and low demand predictability due to erratic end consumer behavior. The need to route large numbers of articles through the suppliers' distribution centers for customer individual picking causes inefficient use of storage space and workforce capacities, as well as excess transport. This chapter outlines a concept on how radio frequency identification (RFID) technology can be used for identification and tracking of individual garment pieces and packages, to allow customer order specific picking already at production plants. This eliminates the need for storage, order-specific picking and packing at warehouses. This chapter describes a prototypical software tool and discusses benefits and requirements for application of RFID technology to support more flexible forms of product delivery to retailers.

Key words: RFID, fashion supply chain, order allocation, direct delivery.

6.1 Introduction

Production and distribution of fashion apparel articles are the last two main steps of an integrated textile process chain. The textile chain comprises all the process steps of the textile and apparel production and distribution process. It also includes manufacture of fibers (yarns and threads) by the fiber industry, and forming of textile surfaces (fabrics), as well as finishing and coloring of surfaces by the textile industry (Abd-Ellatif, 2004; Hurcks, 1993).

Fashion garments belong to the most important consumer goods. During the last two decades, the fashion apparel industry has performed as one of the most dynamic industries in world trade. The geographical distribution of production facilities has moved from Europe and North America to Asia and other developing countries, while main retail markets remain in Europe and North America. About 50% of textiles and over 60% of clothing are produced in developing countries, but over 74% are sold in Europe and North America (Gereffi and Memedovic, 2003). In 2009, global apparel retail achieved revenues of $1032 billion, of which $384.2 billion were achieved in Europe. Global revenues are expected to grow to $1163 billion by 2016 (DataMonitor, 2010).

The European fashion industry as a whole (including textiles, clothing, footwear, leather and fur products, jewellery and accessories) is made up of 850 000 companies, working in the areas of design, manufacturing, distribution and retail of materials and fashion goods. Most of the companies are small or medium-sized. In 2009, the industry employed more than 5 million people, or 3.7% of the European non-financial business economy. Roughly 2 million people were employed in fashion manufacturing, representing 6% of all EU manufacturing jobs. A further 3 million people were employed in related jobs, such as design, branding and marketing, to supply chain management and retail (EC, 2012a). The European textile and clothing industry (not including footwear, leather and fur) was made up of 220 000 companies in 2006, which employed altogether 2.5 million people and generated a turnover of €190 billion. The textile and clothing sector thus accounts for 3% of total manufacturing value added in Europe (EC, 2012b).

Apparel logistics in Europe has to cope with a number of specific problems, resulting from geographically widely distributed supply chains, long lead times, high product variation, short product life-cycles and low demand predictability. This chapter looks specifically from the angle of European fashion apparel suppliers, at how RFID technology can be used to improve the allocation of finished apparel products to customer orders in fashion supply chains. It concentrates on exploring how of RFID technology can be used for identification and tracking of individual garment pieces and packages. Use of RFID allows early allocation of garments to individual customer orders already at the production plants. The consequences this may have for apparel distribution are also discussed.

Section 6.2 outlines production and distribution in the European apparel industry. Section 6.3 describes the differences between warehouse-based distribution of garments and more flexible distribution forms, including direct delivery of the garments. Section 6.4 explains the use of RFID technology for allocation of garments and garment packages to customer orders during delivery of the garments. Section 6.5 describes some aspects of a prototypical software application that has been created to support the described allocation method. Section 6.6 discusses some benefits that can be gained by application of the described method, as well as their requirements. Section 6.7 provides the conclusion with an outlook on future work. Section 6.8 points to additional sources related to the topic of this chapter and Section 6.9 lists the sources used.

6.2 Production and distribution in the European apparel industry

Apparel supply chains are often geographically widely distributed, due to strong outsourcing of fashion manufacturing to low labor cost regions, in particular Asian countries during recent decades. This development is illustrated by the marked increase of international trade of ready-to-wear garments. During the period 1973 to 2000, global exports of ready-to-wear garments grew by more than

1500%, which raised the share of garments on total international trade volume from 2.2 to 3.2% (Grömling and Matthes, 2003).

In this context, European fashion suppliers (or distributors) have reduced, or often completely surrendered, their own manufacturing capacities and have adapted their primary activities towards integration, planning and coordination within geographically widely distributed supply chains (Fissahn, 2001; Langenhorst, 1999).

Thus, as detailed in Table 6.1, typical supply chains include raw material suppliers and garment manufacturers, with production plants situated in Asia as

Table 6.1 Roles of the participating companies within a generic apparel supply chain of the case study

Participant (role)	Location	Tasks, activities
Garment suppliers/ distributors	Europe	Supply and distribution of garments to retailers in Europe and coordination of the garment supply chain. Production of garments either via own production plants, or by subcontract with manufacturers. Operation of one or several distribution centers near customers (retailers) to serve customer orders.
Raw material suppliers	Asia	Supply and delivery of raw materials (dressed fabrics and additional components like zippers, buttons, labels) to production plants.
Manufacturers/ production plants	Asia	Manufacturing and finishing of ready-to-wear garments, either as independent sub-contractors or joint-ventures.
Procurement agencies	Asia	Selection of suppliers and orders of materials needed for manufacture, selection of ready-to-wear manufacturers, price contracting, book keeping and quality control.
Local logistic service provider	Asia	Transport of ready-to-wear garments from production plants to sea or air port.
Global logistic service provider	Asia and Europe	Transport of ready-to-wear garments from Asian sea or air port to European sea or air port.
Local logistic service providers	Europe	Transport of ready-to-wear garments from European sea or air port to distribution center.
Distribution centers (garment supplier)	Europe	Distribution of garments. Picking of garments according to daily customer orders and shipping of garments to the customers. If necessary, taking-back of returning articles not sold by retailers, for temporary storage or for re-selling.
Local logistic service providers	Europe	Transport of the ready-to-wear garments from distribution centers to retailers by lorry or van.
Retailers	Europe	Selling of garments to end users; daily ordering of articles in order to replace articles sold to end-customers and replenish own stocks.

Source: Based on Bruckner and Müller (2003) and Fissahn (2001).

well as local procurement agencies. The suppliers continue to operate ready-made product storage warehouses and distribution centers mainly located in Europe, from which they serve retailers close to the customers. Several different logistics providers bridge the geographical distances and often provide additional services.

Besides the composition of apparel supply chains, other important factors affecting fashion logistics originate from the specific properties of fashion products, which are characterized by high product variation and short product life-cycles, resulting in short product sales periods and low demand predictability (Mostard *et al.*, 2011).

A large number of custom-tailored and non-interchangeable products have to be manufactured for the many existing different supplier or retailer brands, trademarks and product lines, even if many of these products are often relatively similar to each other. Fashion apparel products have to be delivered in many different variants, differing in their colors and sizes (Scholz-Reiter *et al.*, 2009a). The many different products and variants result in large numbers of different stock-keeping units (SKU). Often fashion suppliers have to deal with several thousand, or tens of thousands of different SKU. Small retailers, or outlets, often order garments across a large variety of different SKU, but only a small number of pieces of each SKU (Thomassey, 2010). This still ongoing product diversification induces fashion suppliers to keep pieces of each SKU in stock in order to be able to react faster than the mentioned long delivery times normally allow for. This creates pressure to increase aggregate stock levels (Scholz-Reiter *et al.*, 2009d). Due to relatively high product value per volume, storage areas filled with garment pieces as well as containers filled with garments in transport, represent a comparatively high value for the suppliers as bound capital.

Generally, it is useful to classify apparel products according to their fashion degree, which is inversely related to their life-cycle expectancy. Thus, products can be classified into the following categories:

- Less fashionable, more utility orientated, so-called basic articles (basics), which have longer product life-cycles of several years, during which they are marketed.
- Seasonal fashion products, which are characterized by regular, seasonal exchanges or updates of the product assortment at least twice a year.
- High-fashion products, which are notoriously short-lived, with short marketing and sales periods of only a couple of weeks. Properties of these different classes of apparel products are listed in Table 6.2.

On average, 95% of stock-keeping units change every sales period (season). Thus, they fall into the fashion or high-fashion product categories. Once their regular sales period has ended, many of these seasonal articles can only be sold with large price reductions in discount retail stores, or in less wealthy countries. Demand is more regular for basics, but particularly difficult to predict for short-lived fashion products (Thomassey, 2010), due to the erratic nature of customer decisions based

Table 6.2 Differentiation of apparel products according to fashion degree

Criterion (fashion degree)	Basics	Fashion articles	High-fashion articles
Innovation	Little new product development	Regular new product development	Continuous new product development
Length of life-cycle	Long	Short	Very short
Demand predictability	High	Low	Very low
Sales risk	Low	High	Very high
Focused aspects	Utility is most important	Fashion trend is most important	Fashion trend is most important
Replenishment potential	Replenishment possible	Replenishment is limited	Replenishment is impossible
Production/delivery strategy	Make-to-stock strategies		Make-to-order strategies
ECR application	• Quick Response (QR) • Continuous Replenishment (CR) • Collaborative Planning, Forecasting and Replenishment (CPFR) • Vendor Managed Inventories (VMI) • Consignment Warehouses		• Quick Response (QR) • Collaborative Product Development
Order/delivery models	• Never-out-of-stock (NOS) delivery • (Warehouse stock-based delivery)		• Classical seasonal order business • Key customer order business • (Warehouse stock-based delivery)

Source: Based on Quick *et al.* (2010), Fissahn (2001) and Hoyndorff *et al.* (2010).

on changing fashion tastes (Brun and Castelli, 2008). Delivery dates have to be synchronized with changes between summer and winter collections in stores. The exact dates of the changes depend on seasonal weather changes, which are difficult to predict during production and distribution planning. Retailers, who have to react flexibly to modifications in customer requirements, require the same flexible reaction to demand fluctuation of their suppliers.

6.3 Order and delivery models of apparel products

In reaction to different fashion product types and their different retail or customer requirements, different order and delivery models have been established in apparel supply chains involving suppliers and retailers. The most common order and delivery models are listed in Table 6.2 and their basic structure is illustrated in Fig. 6.1. Selection of the appropriate order and delivery model depends on the

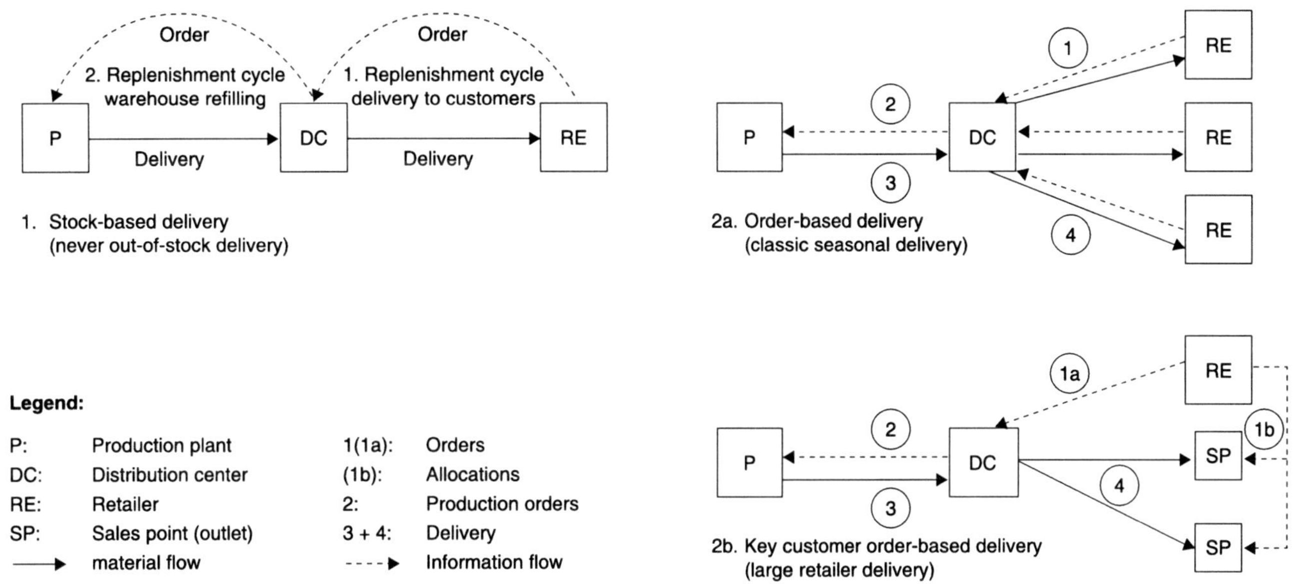

6.1 Order and delivery models in apparel supply chains (source: based on Ahlert and Dieckheuer (2001) and Bruckner and Müller (2003).

longevity of the articles' sales periods and according to customer classes. The duration of the sales periods can be grouped into two classes only: fashion or high-fashion articles sold for only one season and basic articles sold for many seasons. Customer classes can be grouped into two classes: small retailers and large retailers. Small retailers, such as boutiques or millineries, have only one or a few sales points and order small volumes; large retailers, such as clothing chains or mail order wholesalers, often own many sales points and order large volumes.

As is shown in Table 6.2 and in Fig. 6.1, the different order and delivery models can be distinguished according to two basic delivery strategies: make-to-stock and make-to-order. Make-to-stock strategies are largely employed for standard articles, which are sold for several seasonal periods without being modified. Warehouse delivery and never-out-of-stock (NOS) delivery are examples of make-to-stock strategy applications in the apparel industry. Make-to-order strategies are normally employed for more short-lived, seasonal products, which are created and sold only for one season and thus follow recent fashion trends. Classical seasonal business and large retailer (key customer) business are examples of make-to-order strategy applications in the apparel industry.

Most important among make-to-stock delivery strategies is NOS delivery, which may be considered a fashion retail specific form of continuous replenishment, or CR retail strategy (von der Heydt, 1999). NOS delivery is applied to basic products sold over long periods of several years. As illustrated in part 1 of Fig. 6.1, the basic structural characteristic of NOS delivery is synchronization of two replenishment cycles, one short-term for retailers, and one more long-term orientated for apparel suppliers, concerning order, sales and stock volumes. Many retailers only keep base stock inventories of NOS articles, while the suppliers keep inventories to replenish retail stocks with short lead times. Retailers review daily (or, in some cases, weekly) the number of pieces sold of each SKU in each of their sales points and reorder these overnight at the suppliers' warehouses. Delivery times from the warehouse to the customer range from one to three days. Service level agreements (SLA) between supplier and retailer specify the availability of ordered articles within certain boundaries. Supplier warehouses are refilled periodically from production with a more long-term replenishment process (Ahlert and Dieckheuer, 2001).

Another form of stock-based delivery, which is now decreasing in importance, can be described as follows. Suppliers occasionally produce articles based entirely on their own market estimates, without any pre-orders, and then sell them with short delivery times directly from their warehouses, until stocks have run out (Fissahn, 2001). For NOS delivery in particular, suppliers often have to keep large product stocks in order to be able to react to demand fluctuation faster than the already mentioned long delivery times allowed, and thus maintain high service levels.

Classical seasonal business and large retailer (key customer) business are the most important implementations of make-to-order delivery strategies. Classical seasoning business (Ahlert and Dieckheuer, 2001; Fissahn, 2001) is characterized by fixed seasonal cycles with fixed dates, or phases, for product offers, orders and delivery. A supplying apparel company presents its collections of articles to potential customers some six to eight months before regular delivery dates. During a defined pre-order period of several weeks or months, retailers can pre-order pieces of the presented articles in the desired quantities and variants. Delivery of the products is guaranteed either at a fixed date or, in some cases, within a fixed time span around a pre-ordered date. These delivery dates are synchronized with anticipated changes between seasons (e.g. summer and winter clothes) in stores. Production volumes are based on aggregation of the retail pre-orders, though in practice, production planning and procurement of raw materials have to start already before the end of the pre-order period, using forecasts of total demand. Products may also be produced in larger quantities than if only based on pre-orders.

The part of the production volume, which has been produced in excess to pre-ordered volumes, is offered directly from the warehouses to so-called post-orders, until stocks run out (Ahlert and Dieckheuer, 2001). Nevertheless, as illustrated in Fig. 6.1(a), the principal process can be summed up as first collection of retail orders, then placement of production orders based on aggregated pre-order volumes, and at a later date delivery of the ready-made garments to the ordering customers. Depending on the type of products, the described process is repeated several times a year. Each repetition may be called a cyclical process, or cycle. Classical seasonal order is directed largely to small retailers, such as millineries or boutiques. For this reason it often involves a large number of customers. Most of the customers order only a small number of articles belonging to different SKU, often just one or two pieces for each SKU.

As the fashion products expire quickly and can only be sold during a single, short sales period, no re-production of additional volume is possible before the end of the sales period. Thus, production volumes have to be fixed at, or before, the beginning of the sales season. At the end of the sales period, the stock levels of the fashion products should ideally be close to zero, as they cannot be sold with profit in the next season. However, stocks should not run out before the end of the sales period. However, usually production and storage of fashion products cannot be handled efficiently enough for stock levels to be close to zero at the end of a sales season. It is estimated that roughly 30 to 40% of the articles produced for one sales period have to be written off at the end of the sales period, or sold with high price cuts, whereas on the other hand, part of customer demand cannot be satisfied, with corresponding loss of potential sales revenues (Hoyndorff *et al.*, 2010).

For key customers, who regularly order large quantities of articles and thus are important for the supplier's success, for example, clothing chains or mail order wholesalers, the original classical seasonal order business has been modified in certain ways in reaction to these customers' special needs. They often run own brands and trade labels requiring specialized, custom-made products. Thus, product development is often performed as a collaborative process of supplier and retailer. Ordering and delivery are generally not restricted to pre-fixed dates but are largely dependent on customer wishes for delivery dates (Fissahn, 2001). As illustrated in Fig. 6.1(b), orders of large retail chains are placed by the chains' central procurement departments, based on aggregation of the individual demands of their many sales stores and outlets. Garments can be sent by the suppliers directly to the different sales points, or to distribution hubs operated by the retailers. Even in the latter case, the garments are picked and bundled according to the orders of the individual sales points at the suppliers' distribution centers. The individual bundles are then aggregated into larger transport freight loads. As larger retail chains often operate hundreds of outlets or sales points scattered across several different countries, the distribution effort and complexity are similar to classical seasonal business, where a large number of small retailers are the customers.

Several hybrid order and delivery strategies exist, combining different features of the described order and delivery strategies described above. One example is so-called seasonal filling up products, which combine large retailer orders and warehouse-based delivery or NOS delivery of products during a summer or a winter season. For the next half year, existing retailer stocks are exchanged for the complementary season's products and stored at supplier warehouses, to be delivered again, when the same season arrives during the next year.

Different information management and organization-related solutions have been introduced to improve supply chain management and better cope with supply chain and stock control problems. Most important in the context of end consumer-focused industries, such as the apparel industry, are Efficient Consumer Response (ECR) concepts and methods (Seifert, 2002; von der Heydt, 1999), which deal with collaborative demand, supply and stock control between retailers and suppliers using electronic data interchange (EDI) channels. Among ECR approaches, the concept of Quick Response or Rapid Response (Richardson, 1996), which emphasizes quick and efficient reaction to market changes, is most relevant for the fashion industry. A concept known as collaborative planning, forecasting and replenishment (CPFR) (Seifert, 2002) additionally includes joint planning and coordinating future operations, including in particular collaborative product development and production as well as distribution planning. As shown in Table 6.2, ECR concepts are applied in the stock-based delivery model and the order-based delivery model. However, they can be better applied with long-term, or basic products, which are mostly sold from stock.

As make-to-order products are more challenging in apparel logistics, the concepts for distribution in classical seasonal business and large retailer order business will be the focus of the following sections of this chapter.

6.4 Warehouse-based versus flexible garment distribution forms

Currently, distribution of the ready-made garments to the retailers is often based on routing all garments by default through the suppliers' distribution centers, where garments are picked and packed for individual customer orders. At the production plants, the ready-made garments are aggregated into large transport units, and packed and put into containers for transport by ship or aircraft. These transport units are product type and variant specific, or random; however, they are not order-specific, as order-specific picking is restricted to the warehouse. During packing, the garments are counted and the quantities documented in packing lists, which serve as arrival forecasts for distribution centers.

The ready-made garments are transported in standard containers, packed into cartons. A container load consists of 200 to 500 packages of up to 20 pieces each. The distance between the production facilities in Asia and the distribution centers in Europe is normally bridged by ship or, in urgent cases, by more expensive air transport. Container vessel transports often need 30 to 40 days until arrival in Europe (Pfohl *et al.*, 2007), compared to two or three days duration of air transport. Both kinds of transports are often handled by external service providers.

At the distribution centers, incoming containers are unloaded and the packages are counted and compared to dispatch notes. However, at this stage, only small samples of the packages are opened for random checks of the garment pieces. Thus packing or counting errors at production are often detected only during later picking, or even later, by the customers. The packages are stored at free space storing areas. When final delivery is due, picking teams open the packages and take from them the ordered pieces. These are then newly packed and directed to the shipping areas for transport to the customers.

Current process execution with routing of all garment pieces through the warehouse results in a number of inefficiencies concerning usage of storage space in the distribution centers and the capacities of their workforce, as well as unnecessary excess transport:

- *Storage space problems at warehouses*: Arrivals of garments for a product season or a large retailer order cause short, but large peaks in cumulative warehouse stocks and demand large storage areas. These are stored only for a short time until final delivery to the customers. Basic articles delivered via more continuous, stock-based delivery forms occupy the bulk of the storage areas equipped with infrastructure for easy storage, supervision and picking. The more fashion-orientated, short-lived order-based articles, which occupy large storage volumes for a short time are often stored at improvised, additional storage areas without proper storage infrastructure (i.e. storage racks), which makes proper handling and picking of the garments, or keeping track of them, difficult.
- *Sharp workload peaks*: Picking of orders at the warehouse requires large manual efforts for opening of packages, withdrawal of pieces of the appropriate product variants in the required quantities, and repacking them. Due to the difficulties of accurately planning transport times, workload balancing cannot be applied often, and timely picking of all customer orders is impossible with the warehouse workforce. Thus, additional workers, who are not used to the warehouse processes, have to be hired short-term from work agencies. This results in poorer efficiency of work processes and higher risk of errors.

- *Inflated transport distances and transport times concerning certain customer regions*: Routing of garments destined to customers located, e.g. in Eastern or North Eastern Europe or the Balkans, through the warehouses of Western or Central European suppliers, results in additional transports and warehouse storage times and thus in longer product delivery times. Transporting garments through different countries, especially first importing them into the European Union, and then again exporting them to countries outside the European Union, such as Russia, increases efforts for customs handling and product traceability.

In order to avoid these problems, many European suppliers favor restructuring distribution of at least part of the produced garments to more direct or flexible delivery forms. This means that instead of routing all articles by default through the warehouses for picking, at least part of the articles should be picked at the factories and then be delivered to customers either directly from the factories, or be routed through the warehouse, but without the need for unpacking and picking there (Fig. 6.2).

If retailer orders are already individually picked at production, they have to be treated as individual customer orders from that point on during the whole transport process from Asia to Europe and have to be individually routed through the supplier's distribution network, even though large numbers of garments are transported as one freight load (e.g. a standard ISO container load). In particular, direct delivery has to cope with a number of short-term modifications of customer orders at times when the garments have already left production and are in transport:

- *Rerouting of articles for defaulting customers*: In the classical seasonal business, articles are often ordered by small millineries and boutiques, which face a high risk of insolvency. In some cases, up to 10% of the orders have to be kept back, as the ordering customers have defaulted or are in immediate danger of doing so. The articles serving such customer orders have to be redirected to the warehouses, where they may be marketed or, if possible, be directly sent to other customers, who have still open orders.
- *Redistribution between outlets of large retail chains*: Key customers are often large apparel retail chains with many geographically distributed outlets and sales points, which place large master orders aggregating the demand of all outlets. Due to the demand forecasting difficulties, or business policies, these large retailers often delay allocation of the articles of these aggregated orders to the different sales points, or change earlier allocations. For example, allocations may change due to opening of new or takeover of existing outlets, thus increasing overall demand volumes, which require reallocation of articles from existing outlets.

In addition, actually produced article quantities often differ to some degree from ordered quantities, or part of the articles may be damaged or lost during transport. In case of larger differences, more complete reallocations have to be performed between different outlets. For these reasons, the final and definite allocations are made only when ready-made garments have already left factories and are in transport.

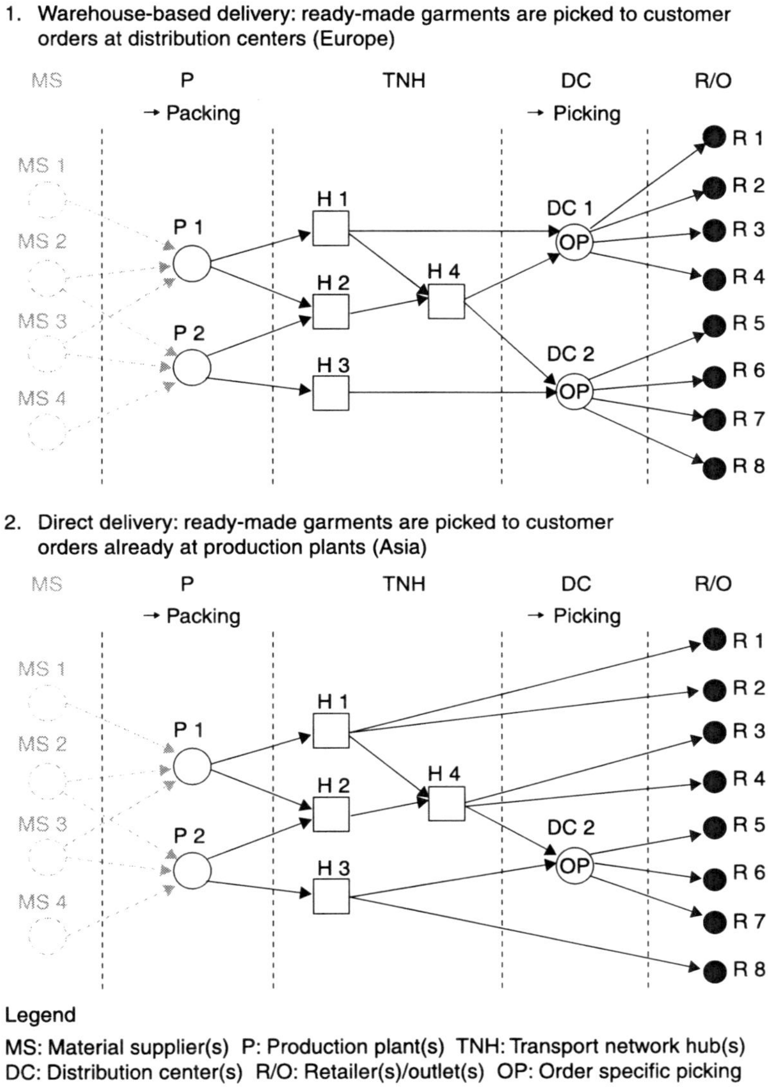

6.2 Customer-specific order picking for (1) warehouse-based delivery and for (2) direct delivery (source: own creation).

With warehouse-based delivery, handling of these modifications or disturbances is not as difficult, as individual articles are allocated to individual customer orders only during picking at the distribution centers, shortly before final delivery to the customers. This allows the apparel suppliers to react to these circumstances, while the articles have arrived there. However, as illustrated in Fig. 6.3, if retailer orders have already been individually picked at production and then have to be individually routed through the supplier's distribution network, these order modifications require more extensive handling. For instance, articles that are in transport to defaulting customers have to be redirected to the distribution center (part 1, Fig. 6.3). Articles on the way to an outlet of a large retail chain, which has modified the distribution of its sales volumes over its outlets, may have to be redirected to another outlet (part 2, Fig. 6.3).

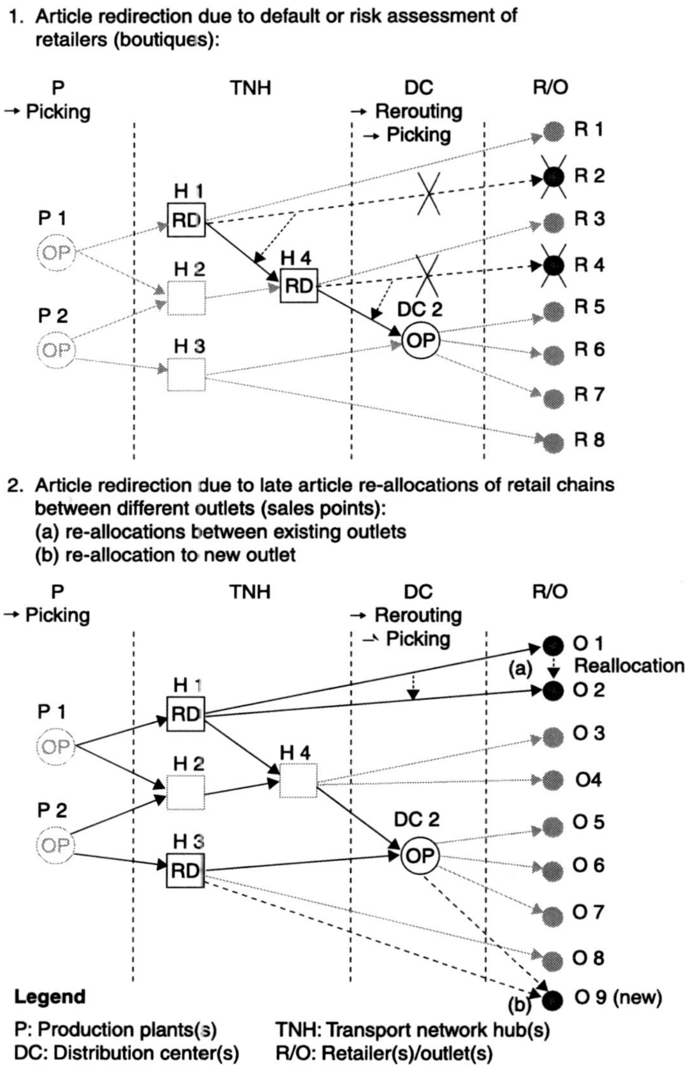

6.3 Redirection of articles due to: (1) defaulting retailers; or (2) re-allocations of articles between different outlets of retail chains (source: own creation).

6.5 Use of RFID technology for garment allocation to fulfil customer orders

The above-mentioned flexible distribution process is established by replacing a conventional process chain with a modified process chain. As described, the conventional process chain consists of the following process steps: garment producing and finishing; non-order-specific packing and transport unit building; transport via transport hubs to distribution centers, storage at distribution centers; order-specific picking and packing; and dispatch to retailers. The modified process chain consists of the following process steps: garment producing and finishing; order-specific picking and packing; transport unit building; and transport via transport hubs. However, if necessary, and possible, packages may be redirected to distribution centers for redistribution processes similar to conventional process execution (storage, re-picking, re-packing and shipping to customers). The precise order of process steps for current as well as modified process chains is listed in Table 6.3. The modified process chain requires that retailer orders as well as garment packages are handled individually during the transport process. In particular, this includes:

- keeping track of individual orders (in classical seasonal business) or 'sub-orders' (meaning the orders of individual outlets or sales points as parts, which are the 'building blocks' of larger, aggregate orders of large retail chains), and their status;
- keeping track of individual garment pieces and of garment bundles (in packages), and their respective status, during the transport and distribution process from the production plant to the customer.
- allocating individual garment pieces or bundles to customer orders or sub-orders, keeping track of these allocations and re-allocating the pieces or bundles to other customer orders, if necessary.

Retail orders specify a number of pieces of one or several SKU (articles characterized by identical type and variant properties, i.e. color and size) to be delivered at a certain date to the ordering customer's address. Orders may include articles of different article types. In particular, orders sent by retailers normally express demand of several articles of the same type, but differing in size, and sometimes in color. For this reason, orders are composed of several homogeneous building blocks, which may be called order lines. Each order line comprises all ordered articles of the same SKU, characterized by identical article type, color and size.

Keeping track of individual garment articles has to start at the production plant, during order-specific picking and packing of the ready-made articles. Optionally, individual garments can already be tracked during production, but this case is not described here. The use of RFID technology in the process will be explained for garments that are transported lying in folded form in carton packages; the exact process flow differs somewhat for garments transported in hanging form.

Table 6.3 Process steps of the conventional, as well as modified, process chains for garment delivery

No.	Location	Warehouse-based delivery	Direct delivery
1	**Production plant**	1a) Manufacturing and finishing of garment pieces	1a) Manufacturing and finishing of garment pieces
		1b) Packing of garments into carton packages; not specific to customer order; packages normally filled with pieces of same product, color and size (SKU)	1b) Order-specific picking of garment pieces and respective packing of garments into carton packages.
		1c) Stuffing of packages into transport containers	1c) Stuffing of packages into transport containers
	Main transport	Transport of containers	Transport of containers
2	**At transport hubs**	**In exceptional cases**	**For part of the packages**
		2a) Re-routing of a whole transport unit (container)	2a) Taking out individual packages from a transport unit (container)
			2b) Re-routing individual packages: separating these packages from the rest
			2c) Building a new transport unit, filling a new container
3	**At distribution center (warehouse)**	**For all packages and pieces**	**Only for redirected packages and pieces**
		3a) Warehouse entry and entry inspection	3a) Warehouse entry and entry inspection
		3b) Storage	3b) Storage
		3c) Package opening, order specific picking and packing	3c) Package opening, order specific picking and packing
		3d) Shipping to retailers	3d) Shipping to retailers
	Final transport	Final transport to retailers	Final transport to retailers

Order-specific picking means taking for each SKU included in the respective customer order out of the available stock of garment pieces the right number of pieces of that SKU, as specified in the order. In order to start picking, a list of individual customer orders is needed. The packers should pick the articles to satisfy the existing retail orders in the sequence as given in the list. Several of the picked garment pieces taken together form a bundle of garment pieces. Normally a bundle will be formed by all garments picked for one retail order. Such a bundle may be packed into one transport package. Alternatively, for a larger order, the bundle may need more than one package, or in the case of very small orders, the bundles for several orders may be transported in one common package.

RFID technology (Finkenzeller, 2004; Shepard, 2005) can be used for individual identification and keeping track of the picked garments during and after packing. The use of RFID technology within the different process steps of a direct delivery process chain is listed in Table 6.4. The allocation of bundles of individually picked garment pieces to packages and then of both to retail orders requires combined use of transponders at item (garment piece) and unit (package) level.

Storage and transport of garments in folded form in packages theoretically allow application of cheap labels, just at unit level, on cartons. This reduces the problems relating to transponder technology and application on or in garments, as well as the number of transponders needed to identify the garments. However, storage at unit level allows only for article type-related data storage, and not for identification of individual garments. For this reason, use of transponder technology for customer order-specific picking and packing also requires the use of transponders at item level. To equip the ready-made garment pieces with transponders at the production plant, custom-made, removable transponder labels can be used, that can be, for instance, printed out at low cost with an RFID capable printer and easily be added to the size books or price stickers attached to the

Table 6.4 Use of RFID technology for direct delivery processes

No.	Location		Direct delivery process step	Use of RFID technology	RFID equipment
1	**Production plant**	1a)	Manufacturing and finishing of garment pieces.	Attachment of RFID labels to garment pieces during garment finishing	RFID labels, RFID capable printer
		1b)	Order-specific picking of garment pieces and respective packing of garments into carton packages.	Identification of individual garment pieces and packages during picking and packing. Creation of a package packing list. Allocation of package to a customer order.	RFID tunnel reader, or handheld RFID reader
		1c)	Stuffing of packages into transport containers.	Identification of garment packages during container stuffing. Creation of a container packing list.	RFID reader gate
	Main transport		Transport of containers.	–	–

(*Continued*)

Table 6.4 Continued

No.	Location		Direct delivery process step	Use of RFID technology	RFID equipment
2	**Transport hubs**		**For part of the packages**		
		2a)	Taking out individual packages from a transport unit (container).	Identification of individual packages, comparison with container packing list and order allocation list.	RFID reader gate
		2b)	Re-routing individual packages. Separating these packages from the rest.	–	–
		2c)	Building a new transport unit. Filling a new container.	Identification of garment packages during container stuffing. Creation of a new container packing list.	RFID reader gate
3	**Ware-houses**		**Only for re-directed packages and pieces**		
		3a)	Warehouse entry and entry inspection.	Identification of individual packages, comparison with container packing list and order allocation list.	RFID reader gate
		3b)	Storage.	Identification of individual packages,	RFID reader gate
		3c)	Package opening, order-specific picking and packing.	Identification of individual garment pieces and of new packages.	RFID tunnel reader, or handheld RFID reader
		3d)	Shipping to retailers.		
	Final transport		Final transport to retailers.	–	–

garments during the finishing process. The transponder attached to a piece of garment has to allow precise identification of the SKU the piece belongs to, including, in particular, the article type (or quality), size and color. An additional serial number of the individual piece is useful, but not necessarily required. During packing of garment pieces into packages, the individual packages also have to be identified. It is useful to equip a package with RFID-coded information on the producer and the production order. As the allocation of pieces or packages to customers should be changeable, neither transponder should contain hard information on the customer, but only a simple serial number for individual identification. An example of coding this numerical information on to transponders attached to garments and packages is provided in Table 6.5.

After packing has been completed, the packages are stuffed into a container, which is used for transport between the production plant and the distribution centers or customers. During container stuffing, the packages may be identified and counted using either mobile or gated RFID readers. An aggregate packing list of all the packages and articles in the container may be automatically created and sent to the target hub. The packing list can also contain information for customs service procedures. Figure 6.4 illustrates a simple example of the allocation of an identified individual package of garment pieces to an individual customer order.

In a transport hub, the processes include warehouse entry, redistribution of packages (as far as necessary) and dispatch. During warehouse entry, an RFID-reader reads the information which is stored on the RFID label attached to each package. The gathered information is stored in the order database. The actual stock level data of the hub is updated; the order database is compared to the packages simultaneously. If the order has been modified or cancelled during transport, the package's destination will be changed to other customers, as described. For instance, a similar order can be brought forward. If the order is still valid, no rearrangement is necessary. RFID readers at the warehouse read the information of every leaving package while it is being placed into a container or trailer. Due to some dynamic influences, as described in Section 6.3, during its transport the customer order that an article bundle or package is allocated to may become invalid, or another more urgent customer order may emerge, which has to be served first. In these cases, at arrival at the hub, a package may have to be redirected to the warehouse, or be re-allocated to another customer order.

In case a package has to be re-allocated to a different customer order, differences between ordered and allocated articles have to be minimal over the ordered product variants. The quantities of each product variant are identified in each package and compared to the ordered quantities. Packages are allocated to those

Table 6.5 Coding of relevant, article and package-related data on article and package transponders

Article identification number				
Example:		1.0160157500.2730.4643.00027		
Check cipher	Article number	Color number	Size number	Serial number
1	0160157500	2730	4643	00027
Package identification number				
Example		2.1531001.71428.0001.0000000		
Check cipher	Producer number	Production order number	Serial number	(fill ciphers)
2	1531001	71428	0001	0000000

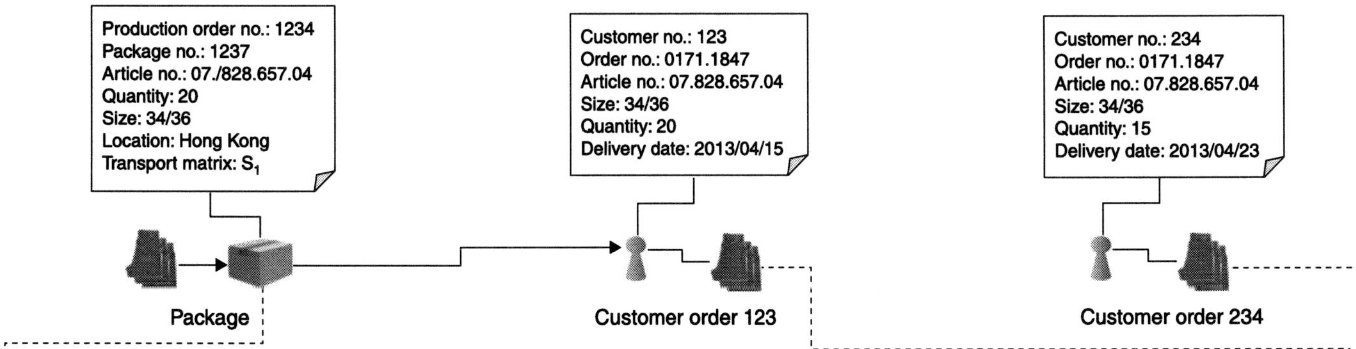

List of identified articles and packages

Package no.	Article no.	Colour	Size	Position	Date	Quantity
1	...	...	...	...	...	...
2	07.828.667.04	60.01	34/32	Hongkong	15.01.2013	20
3	07.828.667.04	60.01	34/34	Hongkong	15.01.2013	20
4	07.828.667.04	60.01	34/34	Hongkong	16.01.2013	20
5	07.828.667.04	60.01	34/34	Hongkong	16.01.2013	20
6	07.828.667.04	60.01	34/36	Hongkong	15.01.2013	20
7	07.828.667.04	60.01	34/36	Hongkong	16.01.2013	20
8	07.828.667.04	60.01	36/34	Hongkong	16.01.2013	20
9	07.828.667.04	60.01	36/34	Hongkong	16.01.2013	20
10	07.828.667.04	60.01	36/36	Hongkong	16.01.2013	20
11	...	...	...	...	...	...

List of messaged orders with their positions

Number	Customer number	Order number	Date of delivery	Article number	Colour	Size	Quantity
1	...	...	...	...	...	...	...
2	123	0171/1847	15.04.2013	07.828.657.04	60.01	32/32	40
3	123	0171/1847	15.04.2013	07.828.657.04	60.01	32/34	40
4	123	0171/1847	15.04.2013	07.828.657.04	60.01	32/36	40
5	123	0171/1847	15.04.2013	07.828.657.04	60.01	34/32	30
6	123	0171/1847	15.04.2013	07.828.657.04	60.01	34/34	40
7	123	0171/1847	15.04.2013	07.828.657.04	60.01	34/36	20
8	123	0171/1847	15.04.2013	07.828.657.04	60.01	36/34	30
9	123	0171/1847	15.04.2013	07.828.657.04	60.01	36/36	30
10	...	...	...	...	...	...	...
11	234	0171/1852	23.04.2013	07.828.657.04	60.01	32/32	20
12	234	0171/1852	23.04.2013	07.828.657.04	60.01	32/34	15
13	234	0171/1852	23.04.2013	07.828.657.04	60.01	32/36	30
14	234	0171/1852	23.04.2013	07.828.657.04	60.01	34/32	30
15	234	0171/1852	23.04.2013	07.828.657.04	60.01	34/34	10
16	234	0171/1852	23.04.2013	07.828.657.04	60.01	34/36	15
17	234	0171/1852	23.04.2013	07.828.657.04	60.01	36/32	20
18	234	0171/1852	23.04.2013	07.828.657.04	60.01	36/34	25
19	...	0171/1852	23.04.2013	07.828.657.04	60.01	36/36	20

Legend

Articles — Package — Customer order — Master data

→ Allocation of package to order - - -→ Allocation within software

6.4 Allocation of a garment package to a customer order (source: own creation).

orders, where the difference between ordered quantities and physical quantities is minimal over all product variants. Several difference measures can be used, such as modulus difference or quadratic difference, which are summed up for each product variant to arrive at an overall difference.

In case the number of articles is insufficient to serve all customer orders, it has to be decided which orders should be served, and which should not. For order ranking in priority, appropriate methods are needed. One potentially viable approach is implementation of rule-based decisions (Böse *et al.*, 2005, 2007). For instance, dispatching rules can be employed, which are widely used in production operations and in distribution operations. Such dispatching rules follow the principle of decentralized control and are very robust in a dynamic flow shop system (Heger *et al.*, 2013). The dispatching rules are sequencing rules designed to route the product or item at local decision points. Many different dispatching rules have been introduced (Holthaus, 1997), of which the most common are First-Come-First-Served (FCFS), Shortest Processing Time (SPT), Earliest Due Date (EDD) and Critical Ratio (CR) rules.

Based on requirements in apparel distribution logistics, the following heuristic rules may be used to prioritize the customer orders for selection by the packages: The shortest servable delivery time rule is an adaptation of the slack time rule used in production control (Kistner and Steven, 2001). Orders are ranked according to the length of the time needed for transport of the articles to the customers. The orders with the shortest times are served first. Alternatively, orders may be ranked by the shortest slack time, or difference between contracted delivery date and anticipated transport times to customer. Application of this rule may decrease total tardiness for customer orders. If customer-based order priority is applied, orders are ranked in priority according to the general importance of the ordering customers for the supplier. For example, long-term customers may be served first. Application of this rule decreases tardiness for orders by large customers, but may increase tardiness for orders by less important customers. Accordingly, this rule is mainly applied in those cases when the shortest servable delivery time rule is indifferent to priorities of two different orders.

The range of potential target destinations incrementally decreases during the production and transport process:

- During production and dispatch of the garments in the factories, the garments can in principle be allocated to all customer orders, which include articles of the same product type and product variants as those produced. Regional distribution of customers is not important at this stage.
- During packing and dispatch of finished garments at the production site, a package's set of potential destinations still includes all customers, who have ordered products of the respective type and variants.
- At a transport network hub, a package's set of potential destinations includes all customers, who have ordered products of the respective type and variants and whose products are routed through the respective hub. Figure 6.5 illustrates

an example of how the potential number of destinations scattered across several European countries, which are delivered from two hubs, is reduced to just the hubs located in France and Spain, after the respective bundle has been routed to the hub located in Northern France.

- Finally, after passing the last transport hub of a package before arrival at a customer, only the finally selected customer order is allocated to each package.

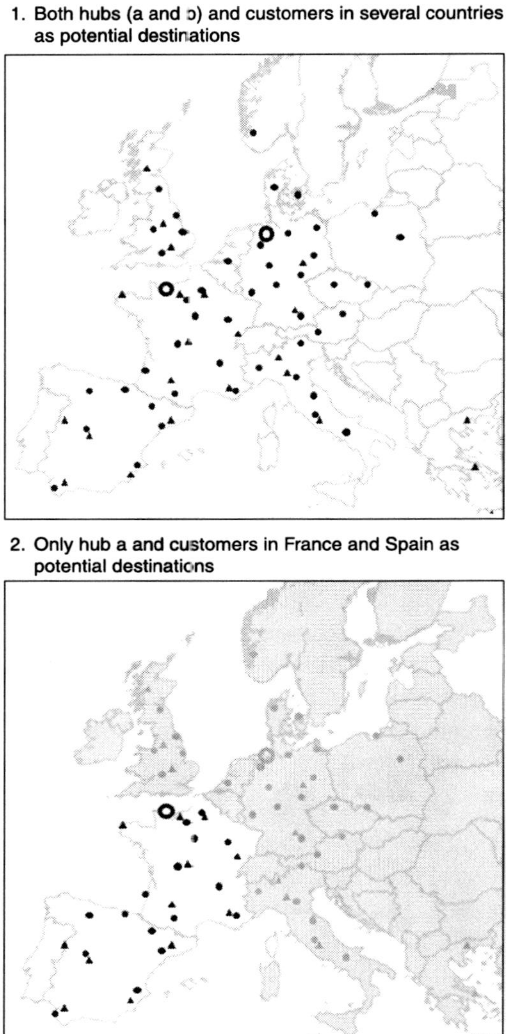

6.5 Decreasing of range of potential target destinations during transport of garments (source: own creation).

The allocation of bundles of garment pieces to individual customer orders is the basis for further decisions during the garment distribution process, such as selection of the next target nodes for transports of garments within the network, and their transport means or storage at the factory.

6.6 Software support for garment allocation

To demonstrate how software can support the described process, an RFID-based test and demonstration tool called *OrdWare* has been created. It supports the use cases *Ordering, Packing, Shipping, Warehouse* (hub) *entry, Storage* and *Palletizing*. Table 6.6 lists the supporting actions provided by the tool for the use cases. The tool offers interfaces for order entry, RFID-based identification of single garments and packages, allocation of packages to orders, and creation of packing lists. Instances of the application can be installed in several network nodes and exchange data in a simple XML format. The architecture of the tool is based on the concept of a network node representing a generalized distribution hub, in which all use cases can be supported (Fig. 6.6). Thus the software supports, at each hub, all use cases listed. Figure 6.7 illustrates exemplarily in the form of a UML class diagram the static structure of the application's *OrdWarePacking* component, which handles the *Packing* use case.

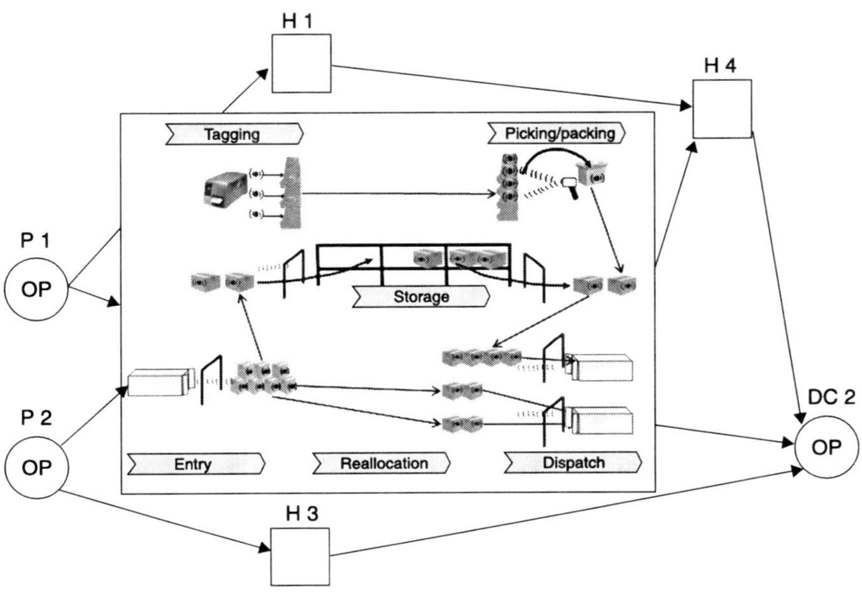

6.6 Network node as basic design structure of *OrdWare* application (source: own creation).

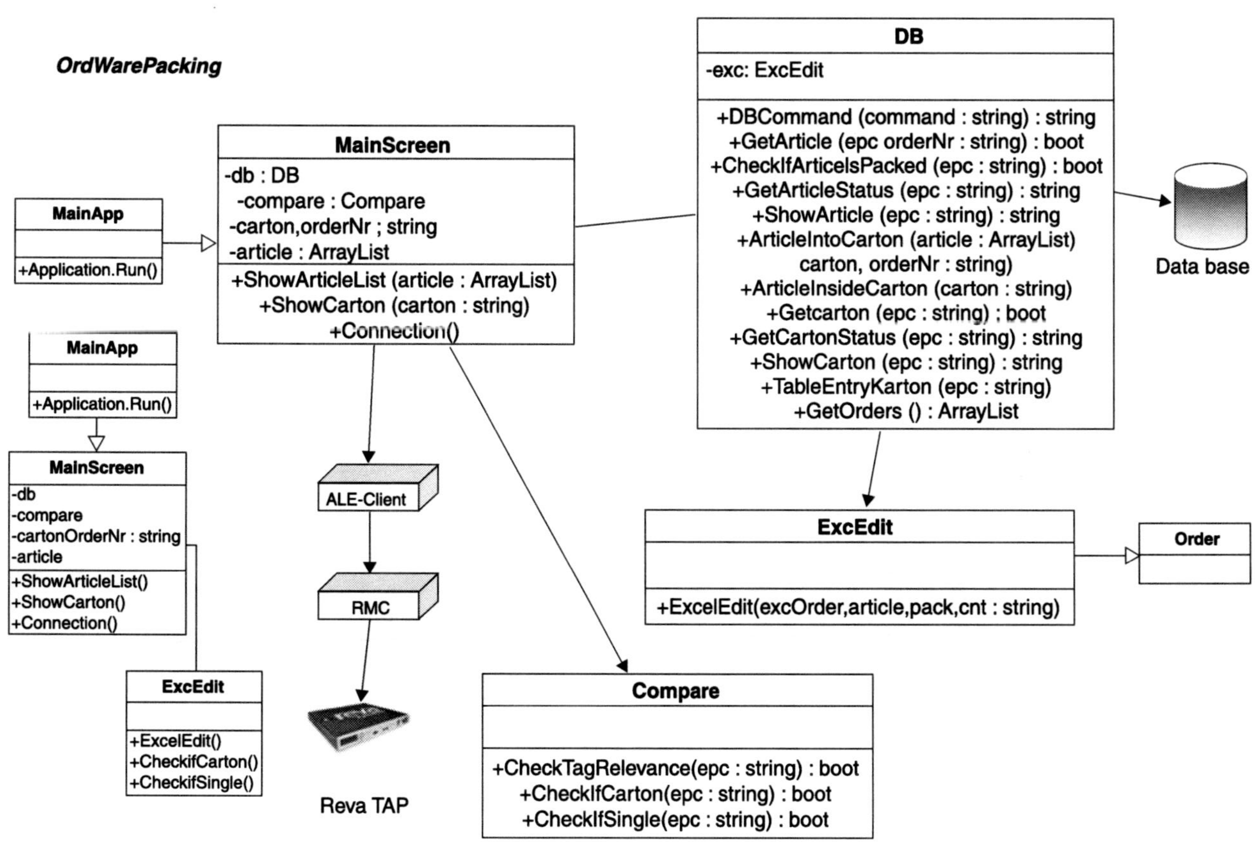

6.7 Static structure (UML class diagram) of the *OrdWare* application components handling the *Packing* use case (source: own creation).

Table 6.6 Use cases supported by the *OrdWare* software tool

Use case	Process steps supported by *OrdWare* tool
Ordering	Messaging and receipt (or entry) of customer orders
Tagging	Attachment of transponders on ready-made garments
Packing	Identification of all individual garment articles and packages by reading of attached RFID transponders Allocation of all garment pieces to package, in which it is packed Allocation of garment packages to customer orders within system's order database
Shipping	Identification of garment packages during container stuffing and creation of an allocation relation Allocation of packages (and pieces inside) to container and transport Creation of a container packing list
Warehouse (hub) Entry	Identification of packages (and inside pieces) after unloading of containers Check with packing list received from sender
Storage	Identification of packages (and inside pieces) during storage in a hub or warehouse Allocation of pieces to storage racks
Palletizing	Identification of packages during palletizing Allocation of packages (and inside pieces) to packages

The class *Compare*, which is shared by all modules of the software, checks the incoming transponder data for relevance. The class *ExcEdit* updates the order lists, which is retrieved from the *DB* class. The *DB* class provides access to the database. The method *GetArticle()* checks whether an article is concordant with the article types and variants of the current order, whether it has been packed already, and whether the ordered number of this article type and variant has already been exceeded. The database and the order list are updated accordingly. The other methods of the class request transponder data from the database, check the status, retrieve article or package data, connect item and package numbers and update the database. The *MainScreen* class accesses the XML data, filters out the tag data, converts them to strings and then further processes them by accessing methods of other classes by way of the *Connection()* method. The retrieved package data is stored as strings; article data is stored as an *ArrayList* for further processing. The components handling the rest of the supported use cases are structured in a similar way.

6.7 The benefits of using RFID technology

RFID technology offers the potential to store individual product-related data and additional logistic process data, both at item and at unit level, allowing the correct

Table 6.7 Application of GS 1 data structure standards

Level of identification	German Standard	International Standard
Identification of transport units (e.g. ISO containers)	Mehrwegtransport-verpackung (MTV) = re-usable unit transport package)	Global Returnable Asset Identifier (GRAI) Global Individual Asset Identifier (GIAI)
Identification of packing units (e.g. palettes, cartons)	Nummer der Versandeinheit (NVE)	Serialized Shipping Container Code (SSCC)
Identification of retail items (e.g. articles)	EAN SGTIN (serialized EAN)	Global Trade Item Number (GTIN)
Identification of a physical, functional or legal unit (e.g. company, or company location, e.g. warehouse)	Serialized Global Location Number (SGLN)	International Location Number (ILN)

Source: Based on Gross (2006) and Koenings (2007).

data to be more easily associated with the products. This allows better synchronization of physical material flows and associated data flows over supply chains (Gillert and Hansen, 2007; Schuster *et al.*, 2007).

The described application of RFID technology and information processing tools allows individual identification and tracking of individual ready-made garments, or of garments bundled in packages, and their allotment to customer orders at different transport hubs, or already at factories. This allows better coping with some problems appearing in particular within geographically widely distributed apparel supply chains with long lead times.

Generally, quality and reliability of logistic services, despite low costs, are crucial for economic success in the apparel branch. Order fulfillment and product delivery failures, such as stock-outs or increased storage costs and product depreciation write-offs, affect revenues in an unfavorable way. It is estimated that, due to the long lead times and demand forecasting difficulties mentioned in Section 6.2, up to 30 to 40% of high-fashion articles cannot be sold during the short sales season before the life expectancy of the article expires (Hoyndorff *et al.*, 2010). The need to cope with this situation, as well as the impetus to realize Quick Response concepts have led in a number of cases to vertical integration of apparel suppliers and retailers by different organizational measures (Richardson, 1996). Vertically integrated suppliers generally achieve better reaction to market changes due to shortened reaction times and more closely integrated information flows. However, these competitive advantages cannot easily be copied by most traditional suppliers, which are often SMEs. It is estimated they have to write off up to 15 to 20% of their high-fashion articles (Hoyndorff *et al.*, 2010). In this context, the described application may help to improve handling of customer orders and thus cope with the following problems:

- *Reduce the risk of stock-outs* caused by incorrect packing lists due to manual counting or documentation errors.
- *Reduce lead times of critical customer orders* by allowing these orders to be served directly, without need for further storage and picking at the distribution centers.
- *Improve the handling of articles produced for extinct customer orders*: The optimal reaction to the process disturbances mentioned in Section 6.3 would consist in simply redirecting the package of garments accordingly. However, due to the large number of different product variants, it is rare that an open customer order will share the exact SKU in the exact number of pieces of each SKU. For this reason, reaction to order modifications will mainly consist in rerouting these packages to a supplier warehouse. But even when these articles cannot be redirected to other customers with still open orders, they can be better identified, thus allowing redirection to the distribution centers and to trigger marketing activities to sell them in a short time. This may help to reduce the volume of articles that have to be written off, or sold with large price reductions, by the end of a sales season.

Use of RFID technology to support the processes described in this chapter has to meet the necessary conditions for successful application of the technology. Some aspects that will be discussed here are application of transponders well suited to garments or textiles, compatibility of the technologies used at the geographically widely distributed hubs of the logistics network, as well as commitment of the different partners within that network to install and maintain the technology, provide the necessary information, and share the related costs.

Use of RFID technology in the apparel industry requires availability of suitable RFID transponders. A variety of transponders have been specially developed to meet the requirements of applications in textiles or garments (Kallmayer *et al.*, 2003; Kern, 2007; Schneider and Gilbert, 2009). Transponders used for textiles or garments can be classified into inlays, labels and tags (Bertolini *et al.*, 2012). Inlays are semi-finished products, composed of an antenna, often made from conductive ink, and a chip placed on a normally very thin and flexible paper, tissue, plastic or thermoplastic support material. They can be used as components of different types of attachable transponders such as labels. Labels are inlays with an added thin support (protective liner, foil or coating), which is normally flexible and can be adhesive. Tags are labels or inlays embodied into a rigid material or support (Bertolini *et al.*, 2012).

Transponders, which are to be permanently applied at item level, are often sewed into the garments during production. They have to be resistant to physical strains, such as heat, pressure and deformation, as well as chemical strains, such as water, and acid or alkaline environments, due to their exposition to washing, dyeing, ironing and folding during production and transport, or during later use by the end consumer. In addition, permanent labels may reduce the wearing comfort of the garments, and generally require acceptance by end consumers. Alternatively, transponders may be applied only temporarily, for example, as removable labels or stickers. Removal of transponders, before handing garments over to the end consumer or the retailer, avoids the mentioned problems. However, it requires additional process steps for removing and recycling, or disposal, of the transponders, and sufficient quality control of the attachment of the transponders to make sure they remain with the garment and are operational during their required operation time. Custom-made, removable transponder labels can be used, which can be printed out at low cost with an RFID capable printer and added to size books or price stickers.

Collecting data from geographically widely scattered logistic objects and disseminating this data in the nodes of the supply network necessitates common standards concerning hardware components, communication protocols, interfaces and data. In apparel logistics, meeting this requirement is rendered even more difficult by the distribution of the supply chain over many countries in different global regions. One solution may be more universal application of the Electronic Product Code (EPC) industrial standard, which has been designed to enable data sharing across enterprises. EPC defines an architecture, which proposes transponders to store only a product identity and use this to reference a data file containing all relevant product-related data via an internet-based registration and access mechanism, enabling participants to obtain a common view of the objects within a business context. EPC (EPCglobal, 2007a) compatible codes, such as EAN or ILE, can be used to specify article or transport unit-related data (Fleisch, 2005; Schuster *et al.*, 2007).

Different coding systems for logistic objects are listed in Table 6.7. Generally, there are two different alternatives for automatic identification of logistic objects. The first alternative is identification of only the type of an object (e.g. an article), without distinguishing between different instances, as with the standardized barcode EAN 128. The second alternative allows additional identification of individual objects, for example, the instances or pieces of the same type, as with SGTIN. Both alternatives are often compatible with each other. For example, SGTIN takes over the typical fields for coding the producing company and the article type from EAN and adds an individual serial number for the individual piece (Floerkemeier, 2005). The so-called EPCIS events (Electronic Product Code – Information Service events) specify an information infrastructure for the standardized exchange of RFID-based data across enterprise boundaries (EPCglobal, 2007b). However, actual implementation of these standards is still incomplete.

Besides the lack of standards for scalable RFID systems on a global scale, distribution of RFID-induced costs and benefits across supply chains is considered an additional obstacle to adoption of the technology (Szmerekovsky and Zhang, 2008). This includes in particular the asymmetric distribution of costs and benefits across logistics. Often costs for collection, retrieval and distribution of data are to a large degree to be taken at the upstream stations of a production network, while the benefits from using this data can be gained at the downstream stations of the production network. To balance costs and benefits evenly across the supply chain, a quantification of both the costs and the value of a particular piece of information in a logistics network is required. However, existing approaches to the cost-benefit-sharing have not been universally successful (Uckelmann, 2012). A potentially suitable approach may be the use of billing systems that allow determination of the costs of generating and messaging a piece of information and fully automated billing of these costs to the recipient of that information (Uckelmann and Scholz-Reiter, 2011).

According to the concepts developed in a collaborative project at the University of Bremen, autonomous control is characterized by interacting elements in non-deterministic, logistic systems, which possess the capability to render their own decisions independently. Application of autonomous control aims to improve the behavior of a logistic system in cases of dynamic instability resulting from sudden changes affecting the system from within, or from the outside. Examples of changes from within are the breakdown of a production or transportation resource. An example of changes from the outside is a sudden change in customer demand. Improved behavior in dynamic situations makes the system more robust (Windt and Huelsmann, 2008).

In logistic processes, intelligent objects should be able to collect and process information on their environments and to identify and evaluate alternative process executions (e.g. alternative transport routes within a logistic network) according to their individual evaluation system (Böse and Windt, 2007), if the concept of autonomous logistic objects is applied to apparel logistics and the allocation of garments to customer orders, article pieces, logistic units (bundles of article pieces) and customer orders. Individual garment pieces may be considered instances of an article type, which specifies potential customers, quality, color and size. Logistic units for storage or transportation bundling together articles that share similar characteristics should be considered autonomous logistic objects and be provided with an identity of their own. Customer orders should also be considered autonomous logistic articles. Smart labels might be considered as the first step to attaching intelligence to the articles, which enables them to render and execute their own decisions referring to their own objectives, such as short lead times until delivery to a customer, or minimizing transport costs (Scholz-Reiter *et al.*, 2010).

6.8 Conclusion and future trends

This chapter outlined a concept of how the use of RFID technology for individual identification of garment pieces, or packages of garments, can be used to support more direct delivery of garments from production plants to retailers. This eliminates the need to route them by default through supplier warehouses for order-specific picking. Direct delivery requires more flexible allocation of articles to customer orders in apparel supply chains, which have to cope with large numbers of different product variants. In this way, the allocation can be made for any garment packages within the supply network, not only at the final distribution center. Allocation decisions can be rule-based, using combinations of priority rules to rank customer orders, and differences across different product variants.

Further research should try to further explore the application of the concept of autonomous control to apparel logistics and refine and expand the model of apparel pieces and bundles as autonomous logistic objects within transport scenarios. Autonomous control might help achieve full flexibility of the product distribution process, when each bundle of articles may render its own decisions, in order to route itself through a transport network from the production plant directly to the customer. The validity of the concept has to be checked by additional test trials or a limited roll-out for individual product lines.

6.9 Sources of further information and advice

Autonomous control of logistic processes has been studied in a large collaborative research project at the University of Bremen. Two publications give extensive information on this collaborative research project, the first by Windt and Huelsmann (2008) on the basic concepts and first applications, and the second by Huelsmann *et al.* (2011) on the limitations of autonomous control. Within this collaborative project, one sub-project has studied the potential for using the concept of autonomous control in apparel logistics in several publications. Scholz-Reiter *et al.* (2009a,b) provide introductions to the topic and outline the basic applications in apparel logistics, while Scholz-Reiter *et al.* (2009c,d) apply the concept of autonomous control also to the realm of apparel manufacturing. Scholz-Reiter *et al.* (2010) have outlined how the concept of autonomous control can be used for flexible product allocation in distribution processes within an apparel supply chain. Scholz-Reiter *et al.* (2011) report some lessons learned on the limitations of autonomous control in practical applications, including apparel logistics, while in Scholz-Reiter *et al.* (2013), a model for cost determination of autonomous control in apparel logistics is described.

Three German language sources deal in detail with many aspects of using RFID technology in apparel distribution logistics: Schmidt and Mannel (2002) report the results of a project involving tagging of leisure wear garments in a factory in Bangladesh; Mannel (2006) describes a process-orientated model for assessment of economic effects in logistics; and Vogeler (2009) describes a detailed implementation procedure for employment of RFID.

6.10 Acknowledgement

The authors are grateful for the generous support received from the German Research Foundation (DFG) as part of the Collaborative Research Centre 637,

Autonomous Cooperation Logistic Processes – A Paradigm Shift and its Limitations (CRC637).

6.11 References

Abd-Ellatif, S. A. M. (2004), *Strukturierungsmethoden als Basis für die Einführung von qualitätsorientierten Leitsystemen in der textilen Prozesskette*. Aachen: University of Aachen. Available from: http://darwin.bth.rwth-aachen.de/opus3/volltexte/2004/996/pdf/Abd-Ellatif_Samar.pdf (accessed 22 January 2013).

Ahlert, D. and Dieckheuer, G. (2001), *Marktorientierte Beschaffung in der Bekleidungsindustrie*, Münster: FATM.

Bertolini, M., Bottani, E., Ferretti, G, Montanari, R. and Volpi, A. (2012), Analysis of the requirements of RFID tags for efficient fashion supply chain management, *International Journal of RF Technology: Research and Applications*, 3(1), 39–65.

Böse, F., Piotrowski, J. and Windt, K. (2005), Selbststeuerung in der Automobil-Logistik, *IndustrieManagement*, 20(4), 37–40.

Böse, F. and Windt, K. (2007), Catalogue of criteria for autonomous control in logistics, in Hülsmann, M. and Windt, K., *Understanding Autonomous Cooperation and Control in Logistics – The Impact on Management, Information and Communication and Material Flow*, Berlin: Springer, 57–72.

Bruckner, A. and Müller, S. (2003), *Supply Chain Management in der Bekleidungsindustrie*, Cologne: Forschungsstelle der Bekleidungsindustrie.

Brun, A. and Castelli, C. (2008), Supply chain strategy in the fashion industry: Developing a portfolio model depending on product, retail channel and brand, *International Journal of Production Economics*, 116, 169–81.

Datamonitor (2010), *Global Apparel Retail*. Available from: www.datamonitor.com (accessed 12 October 2012).

EC (2012a), Policy Options for the Competitiveness of the European Fashion Industries – where manufacturing meets creativity, European Commission Staff Working Document SWD(2012) 284 final/2. Available from: http://ec.europa.eu/enterprise/newsroom/cf/_getdocument.cfm?doc_id=7666. (accessed 2 February 2013).

EC (2012b), Textiles and clothing industry. Available from: http://ec.europa.eu/enterprise/sectors/textiles/index_en.htm (accessed 4 October 2012).

EPCglobal (2007a), The EPCglobal Architecture Framework. EPCglobal Final Version 1.2. Available from: http://www.gs1.org/gsmp/kc/epcglobal/architecture/architecture_1_2-framework-20070910.pdf (accessed 28 March 2012).

EPCglobal (2007b), EPC Information Services (EPCIS) Version 1.0.1 Specification. Available from: http://www.gs1.org/gsmp/kc/epcglobal/epcis/epcis_1_0_1-standard-20070921.pdf, zuletzt aktualisiert am 25.10.2007 (accessed 26 March 2012).

Finkenzeller, K. (2004), *RFID Handbook: Fundamentals and Applications in Contactless Smart Cards and Identification*, Munich: Carl Hanser.

Fissahn, J. (2001), *Marktorientierte Beschaffung in der Bekleidungsindustrie*, Münster: Münster University.

Fleisch, E. and Friedemann, M., *Das Internet der Dinge*, Berlin: Springer, 87–100. DOI: 10.1007/3-540-28299-8.

Floerkemeier, C. (2005), EPC-Technologie: vom Auto-ID Center zu EPCglobal,. In: Fleisch, E. and Friedemann, M., *Das Internet der Dinge*, Berlin: Springer, 87–100.

Gereffi, G. and Memedovic, O. (2003), *The Global Apparel Value Chain: What Prospects for Upgrading by Developing Countries*? Vienna: UNIDO.

Gillert, F. and Hansen, W.-R. (2007), *RFID für die Optimierung von Geschäftsprozessen*, Munich: Hanser.

Grömling, M. and Matthes, J. (2003), *Globalisierung und Strukturwandel der deutschen Textil- und Bekleidungsindustrie*, Cologne: Deutscher Instituts-Verlag.

Gross, S. (2006), *Eine Informationssystem-Architektur für RFID-gestützte logistische Geschäftsprozesse. Fallbeispiele, Konzepte, Handlungsempfehlungen*, Bamberg: Difo-Druck.

Heger, J., Hildebrandt, T. and Scholz-Reiter, B. (2013), Switching dispatching rules with Gaussian processes, *Robust Manufacturing Control*, 73–85.

Heydt, A. v. d. (1999), Efficient Consumer Response – so einfach und doch so schwer. In: Heydt, A v. d., *Efficient Consumer Response – Konzepte – Erfahrungen und Herausforderungen*, Munich: Franz Vahlen GmbH, 2–23.

Holthaus, O. (1997), Efficient dispatching rules for scheduling in a job shop, *Journal of Production Economics*, 48, 87–105.

Hoyndorff, K., Hülsmann, S., Spee, D. and ten Hompel, M. (2010), *Fashion Logistics. Grundlagen über Prozesse und IT entlang der Supply Chain*, Munich: Huss-Verlag GmbH.

Huelsmann, M., Scholz-Reiter, B. and Windt, K. (2011), *Autonomous Cooperation and Control in Logistics: Contributions and Limitations – Theoretical and Practical Perspectives*, Heidelberg: Springer.

Hurcks, K. (1993), *Internationale Beschaffungsstrategien in der Textil- und Bekleidungsindustrie*, Bergisch Gladbach, Cologne: Verlag Josef Eul.

Kallmayer, C., Pisarek, R., Neudeck, A., Cichos, S., Gimpel, S., Aschenbrenner, R. and Reichlt, H. (2003), New assembly technologies for textile transponder systems. In: *Proceedings of the IEEE Components, Packaging, and Manufacturing Technology Society: 53rd Electronic Components & Technology Conference*, Piscataway, NJ: IEEE Service Center, 1123–6.

Kern, C. (2007), *Anwendungen von RFID-Systemen*, Berlin: Springer.

Kistner, K.-P. and Steven, M. (2001), *Produktionsplanung*, Heidelberg: Physica-Verlag.

Koenings, B. (2007), *RFID im Handel: eine geschäftsprozessorientierte Einführung von Transpondertechnologien*, Saarbrücken: VDM-Verlag.

Langenhorst, I. (1999), *Shop-Logistik in der Bekleidungswirtschaft: eine Analyse der Anforderungen herstellerinitiierter Shop-Systeme an die Logistikprozesse der Bekleidungsindustrie*, Frankfurt (M.): Peter Lang – Europäischer Verlag der Wissenschaften.

Mannel, A. (2006), *Prozessorientiertes Modell der ökonomischen Auswirkungen des RFID-Einsatzes in der Logistik*, Dortmund: Dortmund University.

Mostard, J., Teunter, R. and de Koster, R. (2011), Forecasting demand for single-period products: A case study in the apparel industry, *European Journal of Operational Research*, 211(1), 139–47.

Pfohl, H.-C., Gomm, M. and Shen, X. (2007), China: Textil- und Bekleidungs-Supply Chain zwischen Deutschland und China. In: Wolf-Kluthausen, H., *Jahrbuch der Logistik 2007*, Korschenbroich: free beratung, 258–64.

Quick, J., Rinis, M., Schmidt, C. and Walber, B. (2010), *SupplyTex – Erfolgreiches Supply-Management in KMU der Textil- und Bekleidungsindustrie*, Aachen: Forschungsinstitut für Rationalisierung (FIR) e.V. an der Universität Aachen.

Richardson, J. (1996), Vertical integration and rapid response in fashion apparel, *Organization Science*, 7(4), 400–12. Available from (Stable URL): http://www.jstor.org/stable/2635100 (accessed 13 February 2013).

Schmidt, J. and Mannel, A. (2002), *Einsatzpotenziale der Transpondertechnologie in der Bekleidungsindustrie*, Cologne: Forschungsgemeinschaft Bekleidungsindustrie.

Schneider, A. and Gilbert, H. (2009), Textile-Smart-Label (RFID+EAS) für die Bekleidungsindustrie, *Forward Textile Technologies*, 12, 34–36. Available from: http://www.fashiongrouprfid.de/presse_downloads/ftt_Ausgabe0912_TextileSmartLabel_091127.pdf (Accessed 9 February 2013).

Scholz-Reiter, B., Teucke, M., Özsahin, M.-E. and Sowade, S. (2009a), Smart label-supported autonomous supply chain control in the apparel industry. In: *Proceedings of the 5th international Congress on Logistics and SCM Systems (ICLS2009), Program Committee the 5th International Congress on Logistics and SCM Systems*, 44–52.

Scholz-Reiter, B., Teucke, M., Özsahin, M.-E. and Mehrsai, A. (2009b), Autonomous order allocation in large distance apparel industry supply chains based on use of RFID technology (a case study), *Proceedings of the 14th International Symposium on Logistics (ILS 2009)*, Nottingham: Nottingham University (on CD), 329–36.

Scholz-Reiter, B., Sowade, S., Rippel, D., Teucke, M., Özsahin, M. and Hildebrandt, T. (2009c), A contribution to the application of autonomous control in manufacturing, *International Journal of Computing*, 3(3), 279–91.

Scholz-Reiter, B., Teucke, M., Sowade, S., Rippel, D., Özsahin, M.-E. and Hildebrandt, T. (2009d), Applying autonomous control in apparel manufacturing, *Proceedings of the 9th WSEAS International Conference on Robotics, Control and Manufacturing Technology (ROCOM\ '09)*, Stevens Point, WI, WSEAS Press, 73–8.

Scholz-Reiter, B., Teucke, M., Schweizer Virnich, A. and Özsahin, M.-E. (2010), Flexible product allocation in distribution processes in an apparel supply chain, *Proceedings of the International Conference on Logistics and Maritime Systems (LOGMS)*, NN, Pusan, South Korea, 604–13.

Scholz-Reiter, B., Ruthenbeck, C., Teucke, M. and Hoppert, J. (2011), Limitations of autonomous control in practical applications: Report on lessons learned from vehicle an apparel logistics. In: Huelsmann, M., Scholz-Reiter, B. and Windt, K., *Autonomous Cooperation and Control in Logistics*, Heidelberg: Springer, 291–311.

Scholz-Reiter, B., Sowade, S. and Teucke, M. (2013), Modelling the costs of autonomous logistics, *Proceedings of the First Conference of the German Academic Society for Production Engineering (WGP)*, Berlin: Springer, 221–32.

Schuster, E., Allen, S. and Brock, D. (2007), *Global RFID: the Value of the EPCglobal Network for Supply Chain Management*, Berlin: Springer.

Seifert, D. (2002), *Collaboration Planning Forecasting and Replenishment*, Bonn: Galileo Press.

Shepard, S. (2005), *RFID – Radio Frequency Identification*, New York: McGraw-Hill.

Szmerekovsky, J. and Zhang, J. (2008), The effect of supply chain contracts on supplier and retailer costs and benefits in an RFID system. In: Blecker, T. and Huang, G., *RFID in Operations and Supply Chain Management: Research and Applications*, Berlin: Schmidt, 355–75.

Tellkamp, C. and Quiede, U. (2005), Einsatz von RFID in der Bekleidungsindustrie – Ergebnisse eines Pilotprojekts von Kaufhof und Gerry Weber. In: Fleisch, E and Mattern F, *Das Internet der Dinge*, Berlin: Springer, 143–60.

Thomassey, S. (2010), Sales forecasts in clothing industry: The key success factor of the supply chain management, *International Journal Production Economics*, 128(2), 470–83.

Uckelmann, D. (2012), *Quantifying the Value of RFID and the EPCglobal Architecture Framework in Logistics. A Technology-based Approach to Evaluate and Bill Product-related Information in an Internet of Things*, Berlin: Springer.

Uckelmann, D. and Scholz-Reiter, B. (2011), Integrated billing solutions in the internet of things. In: Harrison, M. and Michahelles, F., *Architecting the Internet of Things*, Berlin: Springer.

Vogeler, S. (2009), *Entwicklung eines Vorgehensmodells zur Implementierung der RFID-Technologie in logistischen Systemen am Beispiel der Bekleidungswirtschaft*, Berlin: Universitätsverlag der TU Berlin.

Windt, K. and Hülsmann, M. (2008), *Understanding Autonomous Cooperation and Control in Logistics: The Impact on Management, Information and Communication and Material Flow*, Berlin: Springer.

7

Intelligent apparel product cross-selling using radio frequency identification (RFID) technology for fashion retailing

W. K. WONG and S. Y. S. LEUNG, The Hong Kong Polytechnic University, Hong Kong, Z. X. GUO, Sichuan University, China and Z. H. ZENG and P. Y. MOK, The Hong Kong Polytechnic University, Hong Kong

DOI: 10.1533/9780857098115.159

Abstract: This chapter demonstrates how a combined use of radio frequency identification (RFID) technologies and the Intelligent Product Cross-selling System (IPCS) can improve cross- and up-selling in the retail industry. In this study, two systems have been developed, namely the Smart Dressing System (SDS) enabled by RFID technologies, and the IPCS. The SDS demonstrates a research endeavour in which, unlike the previous studies which focused on transactional data, customers' in-store data can be collected using RFID-enabled SDS. This data can therefore be used for promoting or cross-selling new products to the customers more effectively and efficiently. The IPCS, integrating a rule-based expert system and a fuzzy screening technique, can handle the difficulties of processing linguistic and categorical information. This means fashion designers can recommend appropriate fashion product items for cross-selling with greater ease. The proposed systems' ability to improve selling strategies for the fashion retail industry will in turn help to increase their sales performance.

Key words: information system, multi-criteria decision making, cross-selling, RFID, retailing, fuzzy logic, rule-based expert system.

Note: This chapter is adapted from Chapter 10: Intelligent product cross-selling system in fashion retailing using radio frequency identification (RFID) technology, fuzzy logic and rule-based expert systems, by W. K. Wong, S. Y. S. Leung, Z. X. Guo, Z. H. Zeng and P. Y. Mok, originally published in *Optimizing decision making in the apparel supply chain using artificial intelligence (AI)*, Woodhead Publishing Limited, 2013, ISBN: 978-0-85709-779-8.

7.1 Introduction

The competition in many product markets has increased as the retail market in particular has become saturated. If retailers are to survive, they need to retain existing customers and gain new ones. However, customers tend to switch from one retail store to another in response to more attractive and competitive offers. Competition in mature markets leads to a situation in which each retailer is continually acquiring and losing customers. Retailers are adopting different strategies to retain customers, for example by identifying their purchasing ehaviour through analyzing transaction data. As a result a variety of systems have been developed to cater for the needs of these retailers.

Some of these systems include geographical information systems (Nasirin and Birks, 2003; Tayma and Pol, 1995), inter-organizational information systems (Lin *et al.*, 2003) and data-mining systems (Bose and Mahapatra, 2001; Chen *et al.*, 2008). All of these improve retail decision-making in a variety of areas, including replenishment, inventory control, marketing and promotion strategies. Through the use of various types of database marketing (Kamakura *et al.*, 2003) and data-mining techniques (Hui and Jha, 2000; Lin and Hong, 2008; Padmanabhan and Tuzhilin, 1999), customers' purchasing behaviour can be analyzed and market knowledge can be expanded. However, one common feature of these systems is the reliance on the use of historical transactional data. Although transactional data is an important source of input, no data is available before transactions are made. This chapter reviews research in which customer in-store data is collected and used for promoting or cross-selling products.

Retailers are well aware that, apart from increasing their number of customers, they need to improve profitability through the customers they already have. In other words, increasing the number of transactions per customer will lead to growth, both in terms of profit and customer loyalty. This has led to the use of cross-selling[1] and up-selling strategies[2] (Cohen, 2004; Loeb, 2003). To actualize these strategies, retailers have to make use of historical transaction data to identify customers' preferences. However, historical data can become outdated over time, providing a poor reflection of customers' tastes.

RFID technology enables the collection of data by the reading of tags from a distance, without any need for contact. A RFID system consists of three major components, namely RFID tags (transponders), an antenna and a RFID reader. The latter is usually interfaced to a computer database where product information is held. Tags are attached to product items and when they are brought close to an antenna, they are activated and the product codes are transmitted to the RFID reader. By relating the product codes to the product details stored in the computer database, users can make use of the information for a number of business and management purposes. Many successful RFID applications have been reported over the last few years, particularly in production management (Guo *et al.*, 2009; Yin *et al.*, 2009), supply chain management (Angeles, 2005; Holmstrom *et al.*, 2010; Sarac *et al.*, 2010), and logistics and inventory control (Cakici *et al.*, 2011; Lefebvre *et al.*, 2007; Pei and Klabjan, 2010). Due to the recent push to encourage business owners to adopt RFID technology, RFID cost-benefit analysis has become an important topic.

Many research papers have reported the benefits of using RFID in such applications as material and inventory tracking, as well as logistics in supply chain management (SCM). Jones *et al.* (2005) reported that a faster and more cost-effective SCM system allowed a UK department store to track 3.5 million reusable trays, dollies and cages throughout its refrigerated food supply chain.

This led to a reduction of almost 80% in the time taken to read a stack of multiple trays, thus increasing data accuracy and reliability. Becker *et al.* (2010) proposed a model-based approach for the evaluation of RFID benefits, relating it to business processes. This proved to be very helpful in improving the individual performance measurement of potential RFID investments in an automotive project. Lee and Lee (2010) presented a supply chain RFID investment evaluation model to enhance the understanding of RFID value creation and measurement, as well as ways to maximize the value of RFID technology in a supply chain.

Other areas of RFID application have been explored, including using it to interact with customers and improve retail sales. Although these are comparatively fewer, a small number of research projects relating to the retail business can still be located. For example, Brown and Russell (2007) conducted an exploratory study on the adoption of RFID technology in the South African retail sector and Wamba *et al.* (2008) investigated the impact of RFID technology and the Electronic Product Code (EPC) network on mobile B2B eCommerce in the retail industry. However, these research projects are not on the same scale as those which this chapter describes.

In the fashion retail business, the Prada store in New York seems to be the first documented case of the use of RFID technology to interact with in-store customers (*RFID Journal*, 2002). In this case, the technology was used in the fitting rooms. This was achieved through using a near-range reader to detect the RFID tag on each garment. Therefore garments needed to be at a specific location inside the fitting room, in order that the detection devices concealed behind the walls could read the signals from the tags. Subsequently, production information or a video clip of a model wearing the garment was shown to the customer inside the fitting room. The system seems to have been limited to the fitting room areas, as there was no integration with other areas of the store. For example, dressing mirrors located inside the store were not incorporated into the system. In addition, any information as to the content of what was displayed through the RFID system inside the fitting room was very limited.

Much research can be found on analysis of the transactional sales information in the retail business. However, the number of investigations on the utilization of product information on shop-floor cross-selling, up-selling and sales performance is limited. Utilizing product information to implement cross-selling and up-selling can provide customers with mix-and-match fashion recommendations, in which two fashion items are matched and presented to the customers. Generally, deciding which two fashion items can be matched is only achieved through a person's subjective judgment. They can assess the effect of the mix-and-match performance from multiple perspectives, such as how well the colours of two fashion items match and how their silhouettes fit together. The procedure of fashion mix-and-match can be modelled as a decision-making process, as illustrated in Fig. 7.1.

Thus, the procedure of fashion mix-and-match recommendations is a decision-making process, which involves matching evaluations of a number of

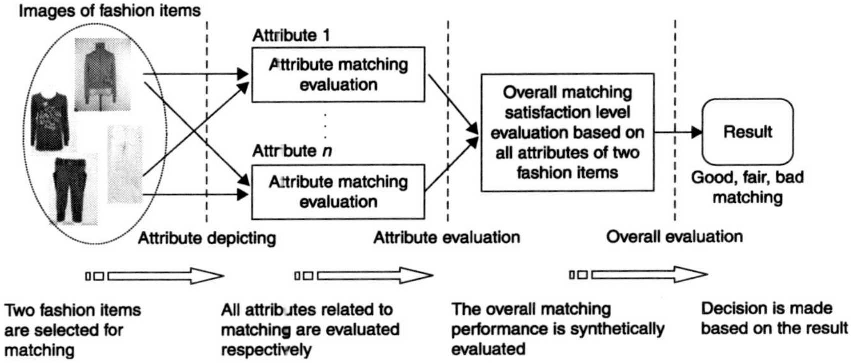

7.1 The fashion mix-and-match decision-making process of fashion designers.

fashion attributes. This is known as multiple criteria decision-making (MCDM). A variety of methods have been developed to solve MCDM problems. The most common methods use mathematical functions to assist decision-makers in constructing their preferences. These functions include multi-attribute value theory (MAVT) (Pictet and Bollinger, 2008; Simpson, 1996), multi-attribute utility theory (MAUT) (Gass, 2005; Khandelwal *et al.*, 2006) and analytical hierarchy process (AHP) (Gass, 2005; Rohacova and Markova, 2009). MAVT and MAUT are difficult to use because the utility elicitation process is time-consuming and complex. AHP is relatively easy to use and less complex than MAVT and MAUT. However, it cannot accommodate the variety of interactions, dependencies and feedback elements which exist between higher- and lower-level elements.

The expert system is a commonly used alternative for MCDM problems (Beynon *et al.*, 2001; Tsiporkova and Boeva, 2006), because it can simulate knowledge base, expertise and therefore the performance of an expert. However, these methods generally assume that all criteria are expressed in precise values. They cannot deal with uncertain, vague or imprecise information. Unfortunately, the decision-making process for fashion mix-and-match involves various imprecise variables, such as the satisfaction levels of different products and the importance level of each attribute. The classic MCDM methods and expert systems are thus not appropriate for the fashion mix-and-match problem.

However, the fuzzy set theory introduced by Zadeh (1965) has been utilized to handle the problem of incorporating imprecise data into a decision framework (Benbernou and Warwick, 2007; Tong, 1982). It can be seen from Fig. 7.1 that the fashion mix-and-match problem mainly consists of two distinct decision-making phases: attribute evaluation and overall evaluation. Due to the complexity and vagueness of these two phases, it is difficult to develop a precise mathematical fashion mix-and-match model. Moreover, some attributes of fashion items are categorical, such as colour, type of apparel, pattern, and so on. These categorical attributes complicate the evaluation process of the fashion designers, because no computing method can be used directly on two categorical values. Therefore, it is infeasible to apply only fuzzy logic methods to the fashion mix-and-match problem, even though there are many ambiguities existing in the overall evaluation. Therefore, the rule-based expert system can effectively handle the large number of rules used for evaluating each fashion attribute, while the fuzzy logic system can handle imprecise variables. As both abilities are necessary in the fashion mix-and-match process, these systems need to be combined.

This chapter demonstrates a combined use of RFID technologies, the expert system and the fuzzy logic method to perform cross- and up-selling, which is implemented by the two systems presented in this chapter. The first system is the Smart Dressing System (SDS), enabled by RFID technologies, which performs several functions including identifying product information, collecting customers' in-store preference, and offering cross- and up-selling (mix-and-match) information. The second system is a hybrid intelligence system, called the Intelligent Product Cross-selling System (IPCS), which has been developed to match customers' selections with other fashion items for mix-and-match purposes. The SDS is therefore the 'front end' which interacts with in-store customers, while the IPCS is the 'back end' responsible for evaluating and matching fashion items. These systems have been implemented in real-life situations, and their potentials and benefits are evaluated in this chapter.

The remainder of this chapter is organized as follows. Section 7.2 introduces the architecture and mechanism of the RFID-enabled SDS. Section 7.3 describes the IPCS and explains how to use the product attributes to generate cross-selling in fashion merchandising. Section 7.4 presents an implementation of the two systems in a fashion retailing company in Hong Kong. Section 7.5 presents the results of this system's evaluation on the retail sales performance. Section 7.6 concludes the study and proposes further work.

7.2 RFID-enabled Smart Dressing System (SDS)

The Smart Dressing System (SDS) was developed in order to achieve a more integrated approach to utilizing fashion product information for cross-selling, and more broadly to explore the potential of RFID in fashion retailing. The design of SDS is based on the current frequency allocation for RFID assigned by The Office of Telecommunication Authority (OFTA) of Hong Kong, this frequency band being 920 to 925 MHz. The system architecture of the RFID-enabled SDS system is shown in Fig. 7 2. The system consists of the following:

- RFID-embedded dressing mirrors;
- smart fitting rooms;
- sales counter with a PC workstation connected to a server;

- RFID reader;
- middleware component inside a retail store;
- smart dressing server;
- product database for cross-selling.

Figure 7.3(a) shows a fashion retail shop equipped with the SDS. All fashion merchandise items in the retail store have an ultra-high frequency RFID tag (Fig. 7.3(b)) attached. The dressing mirror and fitting room are equipped with RFID antennas and projection devices. The RFID antennas are connected to the RFID readers, which are connected to the RFID middleware[3] and smart dressing server. This forms an RFID data management platform to collect, filter and route raw RFID data from individual readers. It avoids network congestion through the automatic data filtering of all the raw data entering the network. Any useless raw RFID data is filtered away to ensure that no 'noisy data' can enter the network, thus improving network efficiency. The RFID antennas are used to detect RFID tags in front of the dressing mirrors and inside the fitting rooms. When an item is brought in front of a dressing mirror or into a fitting room, it can be detected and the antenna will convey the information to the RFID reader as well as the smart dressing server.

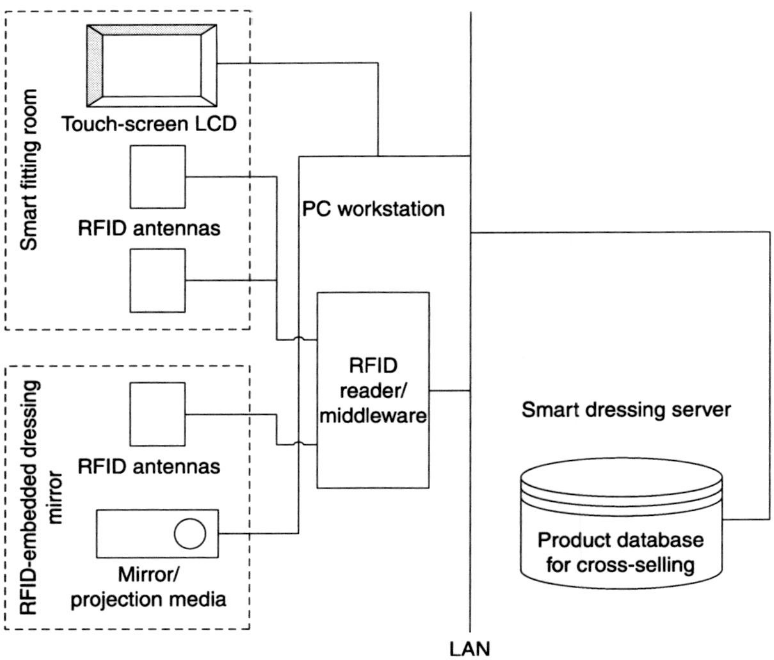

7.2 System architecture of RFID-enabled SDS for cross-selling in the fashion retail industry.

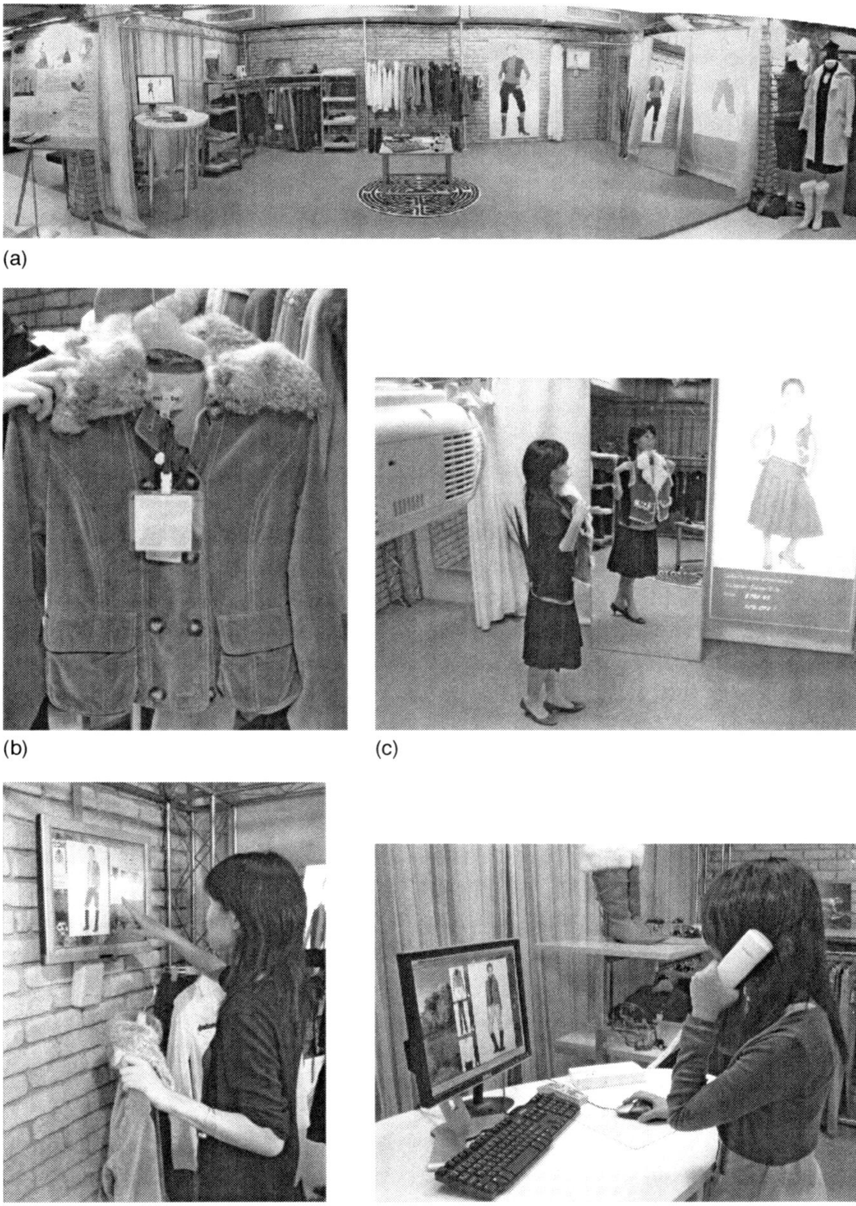

7.3 (a) A fashion retail shop equipped with SDS; (b) a jacket attached with a smart tag; (c) smart dressing mirror displaying mix-and-match items based on the customer's selected item; (d) a customer inside the smart fitting room interacting with the system for more product details; and (e) a salesprson is communicating with a customer inside the fitting room.

The system then delivers cross-selling (mix-and-match) recommendations to the customer, using a human-size display next to the dressing mirror (Fig. 7.3(c)). If the customer is inside a fitting room, the recommendations will be displayed through a touch-screen LCD monitor, which also allows the customer to interact with the system for more product information, for example, colours, sizes, fabrics, etc. (Fig. 7.3(d)). If the customer is interested in the recommended items and wants to try them on, an intercom system is available to inform the sales staff of any clothes that need to be brought over to the customer inside the fitting room (Fig. 7.3(e)).

The store counter, being the control centre of the system, can synchronize with either the smart mirror or the smart fitting room systems. Using the computer and the intercom systems at the store counter, the sales staff can know exactly which items are required by the customer inside the fitting room. They can then take the matching product items to the customer. This eliminates the trouble of the customer having to redress, walk out of the fitting room, select the items, walk back, and try on again.

The store counter is also equipped with a hand-held scanning system whose purpose is to associate a barcode ticket with a smart tag. As many suppliers are not RFID-enabled, fashion merchandise items arriving at a store are only tagged with traditional barcode tickets. The hand-held scanning device reads the barcode and allows the system to generate a smart tag for the item. When a fashion item is sold at the store counter, the smart tag is removed and reused for other items.

7.3 Intelligent Product Cross-selling System (IPCS)

The smart dressing server requires both a front-end and a back-end subsystem. The front-end subsystem, i.e. RFID-enabled SDS, is primarily used to interact with customers and allow the salespersons to provide service to them. The system can display the product details of selected items or recommend new matching fashion items through various display devices. The back-end subsystem, i.e. the IPCS, is designed to assist fashion designers or stylists to streamline the process of making the mix-and-match pairs. This is achieved through comparing the importance of the characteristics (or attributes) of each fashion item with one another.

The system is designed to allow the input of the fashion merchandise characteristics in the form of data, which is then stored in the product database. These characteristics, such as colour, pattern and product type, defined as product attributes throughout this chapter, are used to portray the fashion 'image'. These product attributes are identified through interviews and surveys with the fashion designers. Only those attributes that are important to fashion mix-and-match are identified as the product attributes. Nine product attributes were identified, as shown in Table 7.1. It should be mentioned that although the

Table 7.1 The attributes related to fashion mix-and-match and their level of importance

Attribute	Description	Degree of importance
Type	The categories of apparel items, such as skirt, jacket, etc.	*EH*
Colour	The overall harmonization of the combination of colours in creating the total look	*EH*
Size	The size of apparel	*VH*
Length	The length of apparel to create layers	*H*
Texture	The surface appearance of the materials	*H*
Pattern	The arrangement of design on/in the fabric	*M*
Silhouette	The overall outline or contour of the look in presenting the body figure	*H*
Occasion	The appropriateness of dressing in certain situations or conditions, such as formal, casual, etc.	*VH*
Trend	The general directions that govern what is in or out of the fashion trend	*H*

EH – Extremely high; VH – Very high; H – High; M – Medium

'price' attribute is an important factor in the customer's decision to purchase in fashion retailing, this is not an important factor to fashion mix-and-match and thus all nine identified attributes are related only to the product features. Each day the smart dressing server receives updated mix-and-match recommendations generated from the IPCS. These recommendations are then presented to the customers.

The SDS described in Section 7.2 establishes an environment in which the customer can interact easily with the computer system and obtain the fashion mix-and-match recommendation generated by the IPCS. The SDS is the foundation for cross-selling fashion products and the IPCS is the kernel, evaluating the matching performance of each fashion pair and providing well-matched fashion recommendations to customers.

7.3.1 Overview of the IPCS architecture

Figure 7.4 illustrates the architecture of the IPCS. The IPCS is composed of a rule-based expert system for matching product attribute evaluations, and a fuzzy screening module for generating the final mix-and-match recommendations.

In the rule-based expert system, fashion designers' experience and rules of thumb for matching fashion products were captured and programmed. The rule-based expert system, which included an inference engine, was constructed to evaluate the matching performance of each product attribute and then the overall

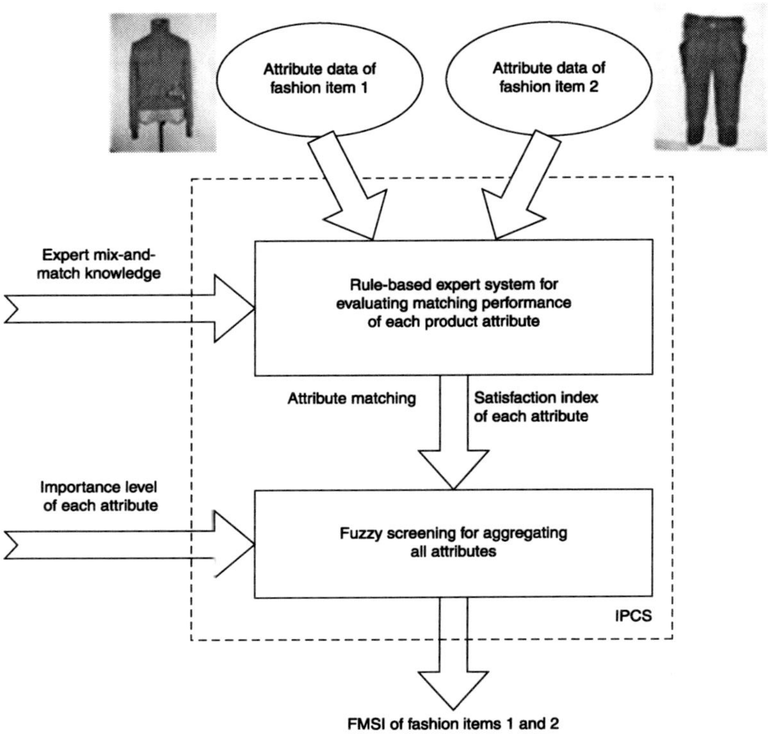

7.4 Architecture of the proposed intelligent product cross-selling system.

performance of the fashion pairs automatically. Since the experience and rules of fashion designers were subjective and involved vague and imprecise information, a fuzzy linguistic rating scale was devised to quantify their opinions. A fuzzy screening technique was used to calculate the overall matching performance of each fashion pair, called the Fashion Matching Satisfaction Index (FMSI). The pairs with the highest FMSIs were used for cross-selling to the customers in the shops. The following section describes the detailed mechanism of the proposed IPCS.

7.3.2 Intelligent product cross-selling based on a rule-based expert system and the fuzzy screening technique

In the IPCS system, there are two phases to evaluate the matching performance of each paired item:

1. *The individual attribute evaluation*: assessing the matching performance of each individual attribute such as colour, pattern and product type of the paired merchandise (outfit).
2. *The overall evaluation*: inferring the overall performance of the pair (outfit) based on the evaluation of all attributes.

These two phases are implemented respectively by using a rule-based expert system and a fuzzy screening technique. One core component of the system is the use of linguistic rating scales to evaluate co-ordinated outfits. This is because it is computationally complicated to evaluate paired fashion merchandise with precise data. In this research, two variables are defined as fuzzy linguistic rating scales. One is the attribute matching satisfaction index (AMSI), denoted as $\tilde{S}$, which is defined as the evaluation of the degree to which attribute matching has been satisfied. This is with respect to each product attribute of the fashion pairs, for example, 'black' colour of a jacket matching 'grey' colour of a pair of trousers. The other variable is the importance index of the product attribute. This is denoted as $\tilde{W}$, and is defined to represent the importance level of each individual attribute in the matching decision. Each fuzzy rating scale can be represented by a number of linguistic terms, called fuzzy numbers.

Before constructing the IPCS system, the two variables $\tilde{S}$ and $\tilde{W}$ were rated by fashion designers on a scale of 0 to 10. With this numerical scale, $\tilde{S}$ and $\tilde{W}$ can be determined by the optimal fuzzy partition using the heuristic cut and trial procedure (Carlsson and Fuller, 1995). Seven fuzzy numbers were used for each of the fuzzy rating scales of $\tilde{S}$ and $\tilde{W}$. Various linguistic terms were defined to represent the rating scales. The linguistic terms of the two fuzzy rating scales with the corresponding fuzzy numbers and membership functions are shown in Table 7.2.

A questionnaire based on the linguistic rating scale (Table 7.2) was then formulated to collect ten fashion designers' opinions about the level of importance of the said nine product attributes (Table 7.1). Almost all attributes in Table 7.1 have the importance level of *H* (high) or above, except the pattern attribute which is rated as *M* (medium).

The rule-based expert system, as shown in Fig. 7.5, consists of a knowledge acquisition module, an expert knowledge base and an inference engine. This is used to compile the attribute data related to apparel and establish rules to conduct attribute evaluation. It has the ability to extract, represent and infer performance rules and patterns. The knowledge acquisition module handles the interaction with experts, converts this knowledge into rules in the format of symbolic representation, and adds or modifies rules for the expert knowledge base. The expert knowledge base is the core component used to store the mix-and-match rules. The inference engine is utilized to emulate the decision-making process of fashion designers, based on mix-and-match rules stored in the expert knowledge base.

Table 7.2 Linguistic terms of fuzzy rating scales with fuzzy number representation

Linguistic terms of the satisfaction degree $\tilde{S}$	Linguistic terms of the level of importance $\bar{W}$	Approximate value of corresponding fuzzy number
Perfect (*P*)	Extremely high (*EH*)	10
Very good (*VG*)	Very high (*VH*)	9
Good (*G*)	High (*H*)	7
Fair (*F*)	Medium (*M*)	5
Slightly bad (*LB*)	Low (*L*)	3
Bad (*B*)	Very low (*VL*)	1
Very bad (*VB*)	None (*N*)	0

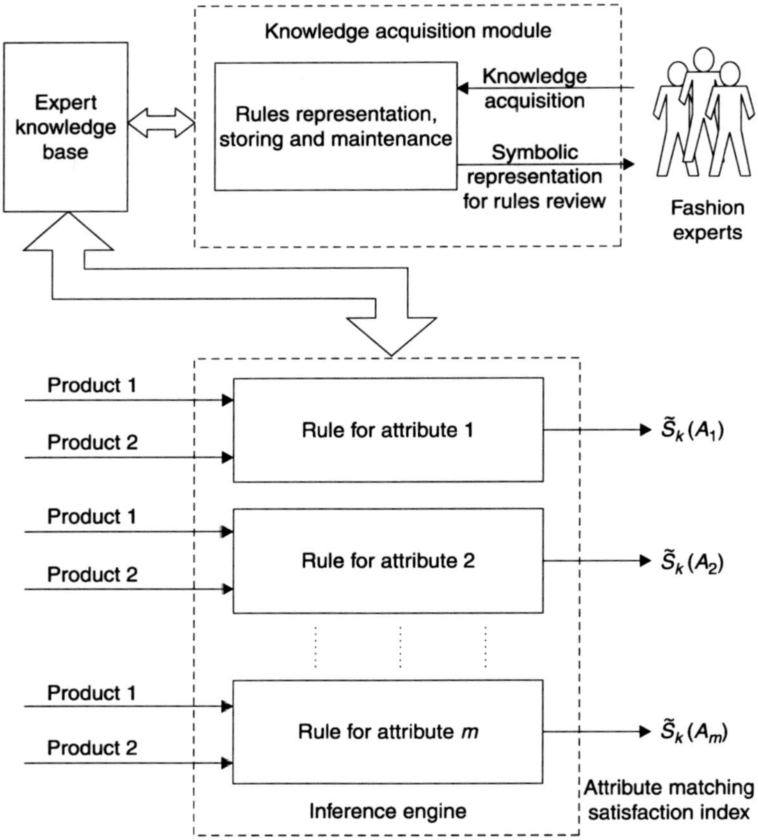

7.5 Rule-based expert system for evaluating the AMSI of each attribute.

The possible values of a product attribute are expressed using descriptive words; for example, the values for colour are black, red, blue, etc. In order to capture the matching rules, the possible values of each attribute should be identified and enumerated. Suppose that a set $A = \{A_1, A_2, \ldots, A_m\}$ denoting the m attributes of apparels is available. All possible values for $A_i (i = 1, \ldots, m)$ are collected and enumerated as $A_i = \{A_{i1}, A_{i2}, \ldots, A_{ib_i}\}$, where b_i is the number of enumerated values for A_i. Therefore, there are $b_i \times (b_i - 1)/2$ rules for attribute A_i. The number of rules for all attributes in sum is $\sum_{i=1}^{m} b_i \times (b_i - 1)/2$.

In the inference engine, the rules of fashion matching performance with respect to an AMSI $\tilde{S}$ could be expressed as an IF-THEN rule of the following form:

> *IF the value of attribute A_i of one fashion item is A_{i1} AND the value of attribute A_i of the other fashion item is A_{i2}, THEN the attribute matching satisfaction index of this paired fashion item in terms of this attribute is $\tilde{S}_k(A_i)$.*

where A_{i1} and A_{i2} are the enumerated values of an attribute, and $\tilde{S}_k(A_i)$ is one term drawn from $\tilde{S}$ (Table 7.2), which is the linguistic rating scale representing the AMSI.

After compiling all the rules, the inference engine is used for evaluation of the satisfaction degree of each product attribute. The inference engine evaluates the matching performance of one fashion pair based on one individual attribute and then aggregates a total matching performance of all nine attributes. This is achieved by using the fuzzy screening technique to calculate the FMSI for each fashion pair.

As there are a large number of possible matches for apparel items, detailed, consistent and precise evaluations of all possible apparel co-ordinations are time-consuming and complicated. The fuzzy screening technique is used to alleviate this problem by binding the space of promising alternatives so that unsatisfactory alternatives can be removed before the detailed evaluations. For a fashion product pair C_j, on the condition that the AMSI and the importance level of each attribute are given, the FMSI of the fashion pair is calculated by the Łukasiewitz implication operator (Carlsson and Fuller, 1995). The detailed procedures of the fuzzy screening technique are described as follows.

Let $C_1, C_2 \ldots, C_n$ denote the n fashion pairs to be evaluated, each pair having m attributes, $A_1, \ldots, A_m$, as criteria for calculating FMSI. Based on the fashion experts' opinion (Table 7.2), the levels of importance for the attributes $A_1, \ldots, A_m$ are determined as follows:

$$\tilde{W}(A) = \{\tilde{W}_{A_1}, \tilde{W}_{A_2}, \ldots, \tilde{W}_{A_m}\} \qquad [7.1]$$

where $\tilde{W}_{Aj}$ pertains to the set of the linguistic rating scale $\tilde{W} = \{\tilde{W}_1, \ldots, \tilde{W}_t\}$, $\tilde{W}_1 < \ldots < \tilde{W}_t$.

First, for fashion pair $C_i(1 \leq i \leq n)$, utilize the rule-based expert system to evaluate the AMSI, and obtain a collection of m satisfaction degree index corresponding to m evaluation criteria, that is:

$$\tilde{S}_i(A) = \{\tilde{S}_{i1}(A_1), \tilde{S}_{i2}(A_2), \ldots, \tilde{S}_{im}(A_m)\} \quad [7.2]$$

where $\tilde{S}_{ij}(A_j)(1 \leq j \leq m)$ is the matching satisfaction index of attribute A_j, which pertains to the set of the linguistic rating scale $\tilde{\mathbf{S}} = \{\tilde{S}_1, \ldots, \tilde{S}_t\}$, $\tilde{S}_1 < \ldots < \tilde{S}_t$ (Table 7.2). Then, according to the fuzzy screening approach, the FMSI, χi, of fashion pair C_i is calculated using the following equation:

$$\chi_i = Min_{j=1,\ldots m}\left\{\tilde{W}_{A_j} \rightarrow \tilde{S}_{ij}(A_j)\right\} \quad [7.3]$$

where $Min(\tilde{S}_i, \tilde{S}_k) = \tilde{S}_i$, if $\tilde{S}_i \leq \tilde{S}_k$, otherwise $Min(\tilde{S}_i, \tilde{S}_k) = \tilde{S}_k$. $\tilde{W}_{Aj} \rightarrow \tilde{S}_{ij}(A_j)$ is defined based on the Łukasiewitz implication operator given by $x \rightarrow y = Min\{1 - x + y, 1\}$, i.e.:

$$\tilde{W}_{A_j} \rightarrow \tilde{S}_{ij}(A_j) = R\text{Imp}(\tilde{W}_{A_j}, \tilde{S}_{ij}(A_j)) = \tilde{S}_{b(ij)} \quad [7.4]$$

$$b(ij) = \begin{cases} t, & p \leq q \\ t - p + q, & p > q \end{cases} \quad [7.5]$$

where t is the number of the linguistic terms, $p(1 \leq p \leq t)$ is the index of $\tilde{W}_{Aj}$ in $\tilde{\mathbf{W}}$ and $q(1 \leq q \leq t)$ is the index of $\tilde{S}_{ij}(A_j)$ in $\tilde{\mathbf{S}}$.

Finally, after the FMSIs of all fashion pairs are calculated using the above method, screen out the fashion pair in which FMSI, χi, is equal to or greater than the predefined minimum satisfaction degree, $\tilde{S}^*$ and recommend it to the customer.

7.3.3 Validation on the IPCS for mix-and-match

To validate the performance of the proposed IPCS system on matching recommendations for cross-selling in a real-life environment, 48 fashion items of a fashion retailer, including a total of 538 expert rules, were used for experimental testing. These samples of fashion items belong to five product types, including dress, jacket, skirt, top and trousers. The distribution of apparel items in each product type and the potential 829 matching pairs are listed in Table 7.3.

To validate the performance of the system, another survey was conducted to collect the matching performance of these 829 pairs rated by 10 fashion designers. We divided the 829 pairs into 7 groups (rated using a 7-level scale) on the basis of their FMSI produced by the program. Table 7.4 indicates the results of the mix-and-match recommendations generated by the IPCS system. The number of pairs in the *P* (*Perfect*) level was zero, which means that the matching rules in the expert system for *perfect* are very hard to meet. The numbers of pairs in the levels of LB, B and VB were 171 (20.63%), 303 (36.55%) and 205 (24.73%), respectively.

Table 7.3 The number of fashion items in each type and their potential matching pairs

Total samples	The number of apparel items belong to each type					Potential matching pairs
	Dress	Jacket	Skirt	Top	Trousers	
48	2	11	3	16	16	829

Table 7.4 Fashion matching results generated by IPCS system

	The FMSI level							Total
	P	*VG*	*G*	*F*	*LB*	*B*	*VB*	
Number of pairs	0	54	32	64	171	303	205	829
Ratio (%)	0	6.51	3.86	7.72	20.63	36.55	24.73	100

Note: *P* – Perfect; *VG* – Very good; *G* – Good; *F* – Fair; *LB* – Slightly bad; *B* – Bad; and *VB* – Very bad.

Table 7.5 Comparison between the result advised by the IPCS system and the evaluation result provided by the fashion designers

		The FMSI level						Total
		VG	*G*	*F*	*LB*	*B*	*VB*	
Expert system		54	32	64	171	303	205	829
Designers	*Same result*[a]	45	26	60	160	289	200	780
	Different result[b]	9	6	4	11	14	5	49
Correct ratio of the expert system (%)		83.33	81.25	93.75	93.57	95.38	97.56	94.09

[a] *Same result* means that the FMSI index of the pair evaluated by the fashion designers is same as the results generated by the IPCS system. [b] *Different result* means that the FMSI index of the pair evaluated by the fashion designers is different from the results generated by the IPCS system.

Each group was assigned to two fashion designers for evaluation. A detailed comparison between the results generated by the proposed system and those provided by the fashion designers is summarized in Table 7.5.

In Table 7.5, it can be seen that the lowest percentage of the correct recommendations conducted by the system was 81.25 and the overall percentage was 94.09, indicating a high level of system accuracy meeting the practical requirements of the fashion retailer.

Table 7.6 The screening performance of the IPCS system at the screening levels of VG, G and F

		The screening level S^*		
		VG	*G*	*F*
Designers	Recommended pairs	50	82	150
	Recommended pairs	54	86	150
IPCS System	Correct recommended pairs (%)	47(87.04)	79(91.86)	149(99.33)
	Missed recommended pairs (%)	3(5.56)	3(3.49)	1(0.67)
	Extra recommended pairs (%)	7(12.96)	7(8.14)	1(0.67)

It can be seen from Table 7.6 that the performance of the system is very satisfactory. The system can achieve 87.04%, 91.86% and 99.33% accuracy when the minimum level of recommendation performance is set at level *VG*, *G* and *F*, respectively. The number of pairs recommended by the fashion designers, which were not recommended by the proposed system, called 'missed pairs', were only 3, 3 and 1 respectively at three different screening levels, while the number of pairs recommended by the expert system, which were not recommended by the fashion designers, called 'extra pairs', were 7, 7 and 1 respectively at three different screening levels. The apparel pairs with very good ratings were exported to the product database of the SDS for cross-selling.

7.4 Implementation of the RFID-enabled SDS and IPCS

Based on the operation logic of the system, the layered technique for software design was employed (Seiter *et al.*, 2000) to develop the system architecture in which there are three layers, namely the user interface layer, the application logic layer and the data layer. The user interface layer receives commands from the users, transfers them to the application logic layer and returns results to the users. The application logic layer is composed of multiple logic processing functions. The object-oriented technique is employed and logic functions are developed module by module. The data layer, where there is a database management system using SQL Server, is responsible for data reading and writing.

Based on the theoretical architectures shown in Figs 7.2 and 7.4, the SDS and the IPCS were developed and implemented in a fashion retailing company, which has nine chain stores in Hong Kong. The product information and the subjective evaluation information of fashion items were input into the system via the back-end subsystem. Figures 7.6(a) and (b) show the interfaces involving the product information, i.e. attributes or features with the corresponding level of importance. Users can use this interface to input the product features such as colour, pattern,

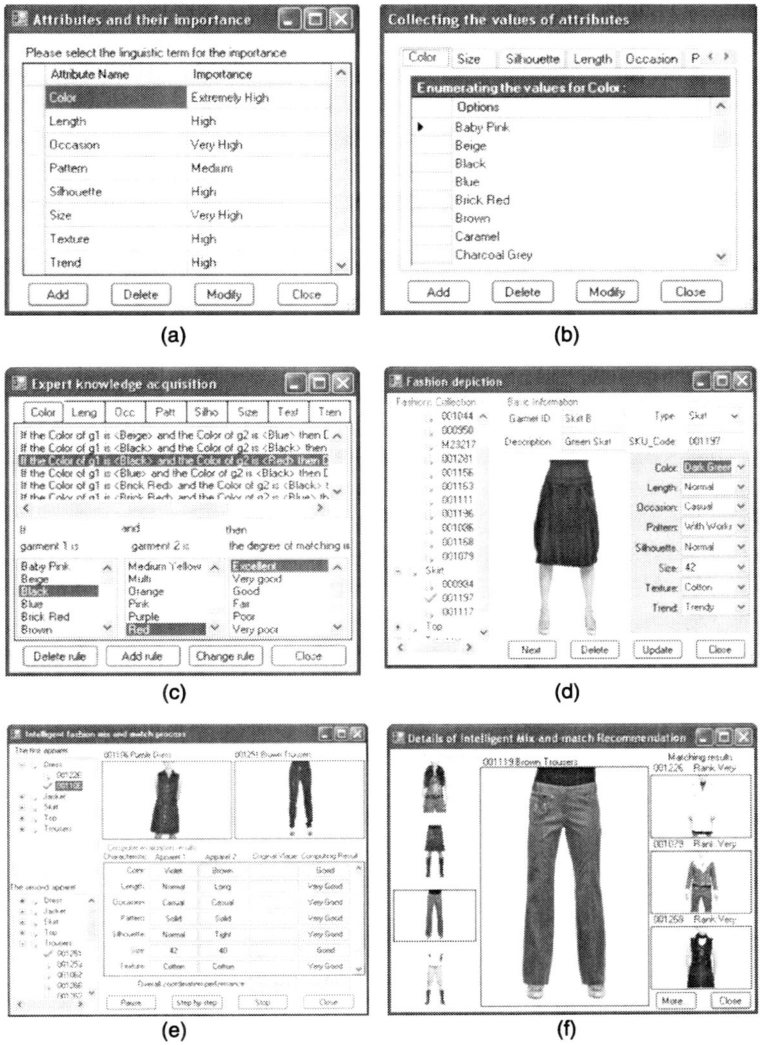

7.6 The client interfaces of the intelligent product cross-selling system: (a) Fashion attributes with their importance level setting; (b) enumerating the values of fashion attributes; (c) storage of expert knowledge acquisition; (d) pairing product attributes for matching performance evaluation; (e) intelligent fashion mix and match processing; and (f) results of fashion mix and match recommendation.

silhouette, etc. with their corresponding level of importance, such as extremely high, very high, etc., which should be considered when evaluating the matching performance of fashion product items. Figure 7.6(c) is the interface used to store the acquired knowledge and rules of fashion product mix-and-match using the fuzzy screening approach described in Section 3.2. Figure 7.6(d) is the interface for the user to input the product attributes of the fashion merchandises which will be paired for matching performance evaluation. Based on the above input parameters, the product items for cross-selling can be generated automatically.

In Fig. 7.6(e), a long vest in brown matched with a pair of pants, also in brown, is evaluated by the system automatically. The matching performance result of each attribute, such as colour, length, etc., as well as overall matching performance, are shown in this figure. Figure 7.6(f) illustrates the recommended paired items which can be used for cross-selling. The smart dressing server receives the updated mix-and-match recommendations generated from the IPCS and these recommendations are then transmitted to the client PC workstations for presentation to the customers on the retail shop floor.

In Fig. 7.7, a pink jacket is selected and brought by a customer to the fitting room. As the jacket is attached with a RFID tag, which is identified by the RFID-enabled SDS, a pair of pants is immediately recommended by the IPCS through a LCD monitor. To make the paired apparel more attractive to the customer, the fashion retailing company utilizes the IPCS by obtaining the matching result first and then arranging a fashion model to dress up in the recommended pair for photo shooting. Figure 7.8 illustrates the mix-and-match recommendations generated by the IPCS through the smart fitting room and smart dressing mirror in the fashion chain store in Hong Kong. Figure 7.9 depicts a recently developed dressing mirror,

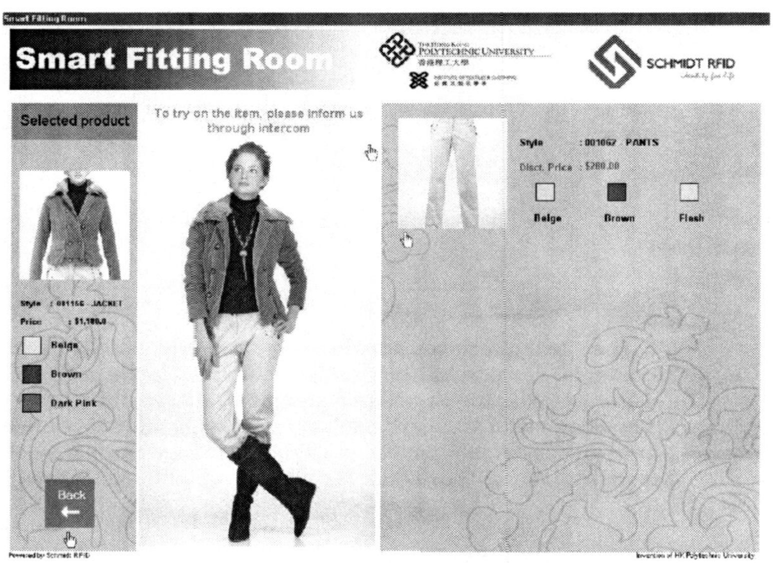

7.7 Mix-and-match recommendations in the fitting room by the smart dressing system.

(a) (b)

7.8 (a) Smart fitting room; and (b) smart dressing mirror, integrated with the proposed IPCS in a fashion chain store in Hong Kong.

7.9 A movable smart dressing mirror.

which can be movable such that no renovation is required to embed the RFID antennas and accessories in the retail store.

7.5 Evaluation of the RFID-enabled SDS

The RFID-enabled SDS, being publicized by the media, was received by the fashion retail sector and developers, and the system was implemented in two of the stores of a fashion retailing company who runs the nine chain stores located in major shopping malls in Hong Kong. In these stores, the fashion merchandise was tagged with RFID, one ordinary dressing mirror and two out of the four fitting rooms were converted into smart devices using the RFID technology integrated with an IPCS for offering mix-and-match recommendations to customers.

Although the company reported that there was an improvement in sales of about 20% a few months after the installation, it is imperative to understand the contribution of the system towards the overall sales. In order to identify the sales improvement brought by the cross-selling function of the system, three-month point-of-sales (POS) data after the installation were collected to evaluate the impact of the system on the sales performance. As customers were not asked or interviewed as to whether or not the purchase was influenced by the system, an alternative approach was adopted: POS data items detected by the antenna within 30 minutes, were counted as a successful application of the system.

Figure 7.10 shows the total sales (ordinary sales and those initiated by the system) of the first three months after the installation. The upper curve indicates the weekly change in total sales quantity and the lower curve provides the sold items, which were detected within the 30-minute limit. On average, the total sales remain more or less the same but the sales bought by the system (lower curve) rises in the last two months. This rise is shown more clearly in Fig. 7.11, which illustrates the proportion of the total sales brought by the cross-selling function embedded in the dressing mirror and two fitting rooms of the system. In Fig. 7.11, there are two trend lines showing a two-stage rise in the system contributing to the total sales; in the initial period (i.e. first month after system installation), about 11% of the total sales were the result of the system, while in the last two months, this contribution improved to over 20%.

The difference is caused by training of the sales staff, although initial training was provided to all staff during the installation and the system was allowed to run for one month. On-site observations found that few customers utilized the system and explored further details of the products, and sales staff were reluctant to encourage customers to use the system for cross-selling. It seems that they were not confident enough to rely on the technology for customer service. Management offered further training to familiarize the staff with the potential of the system and help them understand that the system was to assist their work instead of making staff redundant. There was an obvious sales improvement, in which more than 20% of sales were the result of the system.

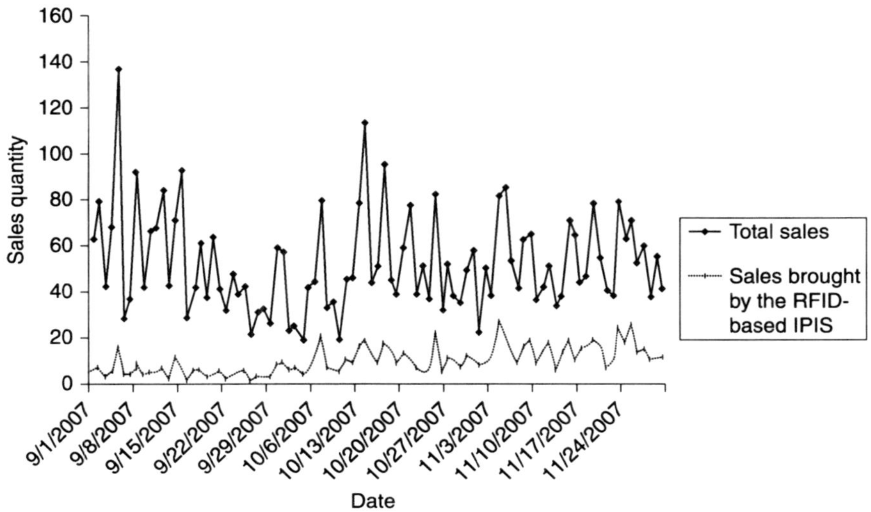

7.10 Comparison of the total sales with sales brought about by the installation of the RFID-based IPCS.

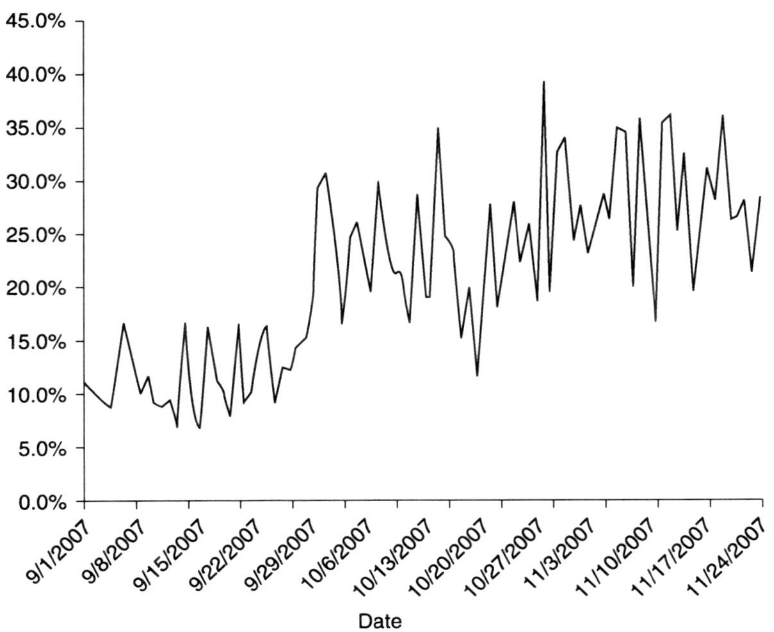

7.11 Proportion of the total sales brought about by the RFID-based IPCS.

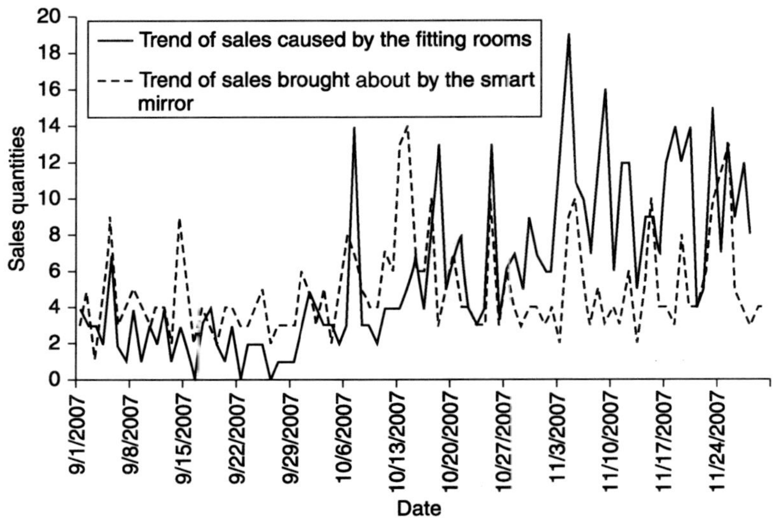

7.12 Trend of sales quantities caused by the two fitting rooms and the dressing mirror embedded with RFID-based IPCS.

As the cross-selling system is embedded in both the dressing mirror and the fitting rooms, it is also important to identify how they contribute to the sales individually. Figure 7.12 illustrates the breakdown of the sales quantities caused by the two fitting rooms and the dressing mirror as a result of the use of IPCS. In the first few months, it was found that the sales generated by the dressing mirror were greater than those by the fitting rooms. This is due to the fact that the human-sized visual display incorporated with the dressing mirror was more impressive to the customer, and the cross-selling recommendations were easily recognized without the assistance of sales staff. Unlike the dressing mirror, in the fitting rooms, the visual display through LCD monitors was comparatively smaller and customers did not feel compelled to use them if they were not encouraged and introduced by sales staff.

After re-training was provided to the sales staff, emphasizing the importance of letting the customers know the mix-and-match recommendations from the fitting rooms, the sales based on the fitting rooms were then greater than those based on the dressing mirror. It was found that the effectiveness and practicality of the system were also determined by the way in which the sales staff made use of the system in cross-selling activities. When sales staff perceive the system as their helping hands and use it effectively, they can recommend customers to have more try-ons based on their selected items.

When the research was carried out, there were a total of five dressing mirrors and four fitting rooms in each store, but only one mirror and two fitting rooms had been installed with the system. It is envisaged that more sales will result if all of them are equipped with the proposed system.

7.6 Assessing the role of RFID technology in fashion retailing

The more conventional use of RFID technologies is in the identification of products in logistic operations; however, the research presented the use of RFID in cross- and up-selling of fashion items. This new approach can bring many benefits to fashion retailers in terms of business performance and customer services.

7.6.1 Early detection of customer's preference

Metropolitan fashion retailers usually know little about their customers in the conventional retailing process. This process usually comprises the following stages: 'walk-in', 'browse', 'fit' and 'go'. Usually the only piece of information a fashion retailer can collect from customers is the transaction and credit card details before customers 'go' out of the store. With so little information to rely on, it is difficult for retailers to enhance their customer service/relationship. The SDS offers service to customers at the earlier stages of 'browse' and 'fit', even before a transaction is actualized. The very action of selecting one or several garments during the 'browse' (or selection) stage will actually 'reveal' the preference of a customer. The SDS system allows the retailer to meet this immediate demand. A customer may be attracted to a style by its colour, fabric texture or the look. By the time the customer removes the item from a store rack and brings it to a dressing mirror for assessing or fitting, the mirror detects the item and shows how the style is dressed up on a human-sized display. It also recommends styles which can mix-and-match with the one the customer is holding. Feedback from the stores using the system showed that customers bought not only their selected items but also the other items displayed and cross-sold on the screen.

7.6.2 Enriching customers' in-store purchase experience without changing the fitting process

The SDS system makes the fitting process more natural because it uses RFID technologies composing predominantly of ultra-high frequency (UHF) tags and readers. These enable the detection of the selected items from a distance and so it is not necessary to bring the clothing item close to any reader. Generally, the signals, unlike using a barcode system where line of sight is required, can be read during the try-on process. There is no change in the way in which a customer fits a piece of clothing, but the new approach allows retailers to have a computerized stylist advising their customers to do mix-and-match.

7.6.3 Cross- and up-selling can be made possible using the smart fitting rooms

Customers who use the fitting rooms have been attracted by their selections (e.g. colours, fabric, style, etc.) and they want to determine whether or not they fit well into these outfits. Detailed product mix-and-match information, together with prices and discounts, can be offered to the customers at this stage. The smart fitting room is equipped with the same technology as the smart fitting mirror, but in order to adapt to the small environment of a fitting room, the human-size display is replaced by a smaller interactive LCD touch-screen display (which is about the size of a desktop display). Thumbnail figures of the clothing items brought inside the fitting room are detected and shown on the screen. By touching appropriate figures, the system provides more information about the product details including available colours, sizes, prices, and even promotions and discounts. At the same time, the screen recommends other styles that can go with the selected item, in the same way as the smart dressing mirror outside.

7.6.4 New fitting and transaction experience

The system synchronizes with the system at the sales counter, so staff can serve the customers with the matching clothing that the customer has not brought into the fitting room in the first place. The customer inside a fitting room can avoid the hassle of putting on his/her own clothes and searching for the matching items from the store racks. In the final transaction stage, fashion retailers could utilize the RFID tags to replace barcodes at the till to generate invoices and finalize fund transfers. This could reduce the time customers wait in a queue to pay.

7.6.5 Standardized cross- and up-selling approach

The output of this research, SDS and IPCS, has enabled a more standardized approach to cross- and up-selling of fashion items: the pre-determined ideas of the company's buyers or designers on how one fashion item matches with another, or how several items are sold as co-ordinates, can be conveyed to sales staff who are the first line in offering services, including styling advice to customers. Customers may turn away due to such reasons as poor service or training of the sales staff; with this system, it will be possible to ensure that cross- and up-selling are executed as originally planned.

7.7 Conclusion

This chapter has presented the architecture of an intelligent system integrated with RFID technology for cross-selling activities in the fashion retail business.

Two systems have been developed and applied in real-life situations. The first system is the SDS, enabled by RFID technologies, which performs several functions including identifying product information, collecting customers' in-store preference, and offering cross- and up-selling (mix-and-match) recommendations. The second system is a hybrid intelligent system, called IPCS, developed to match customers' selections with other fashion items for mix-and-match purposes.

The success of the cross-selling capability of this new dress fitting system relies heavily on the back-end, which is support by incorporating not only the hardware but also the management involved in putting the right styles together, based on fashion design/styling expertise and subsequent modelling and photo shoots. With the collaboration of the company in the case study and fashion design experts in the authors' institution, this research developed the IPCS, which is a system incorporating expert knowledge to carry out initial mix-and-match of the items before a season starts. The IPCS enables the design/styling team to focus on final judging and fine-tuning the mix-and-match outcomes, shortening the time needed for the initial matching.

Utilizing product information to implement cross-selling is a multi-criteria decision-making (MCDM) problem. However, the decision-making process for fashion mix-and-match involves various imprecise variables, such as the matching satisfaction levels of different products and the level of importance of each attribute, which cannot be solved directly by the ordinary rule-based expert systems, since they cannot deal with the problems of uncertain, vague and imprecise information. The attributes of apparel items are categorical but the values of the attributes are literal. For example, the values for colour may be black, red, white, etc. With the 'IF-THEN' rules base of the inference engine, the IPCS, combining the fuzzy screening technique, specifically handles vague and imprecise information in the process of fashion mix-and-match. This application of the knowledge-based attribute evaluation expert system provides an innovative approach to processing linguistic information for similar decision-making problems.

After rolling out the first prototype, further enhancement and commercialization have been licensed to developers. This gives room to the researchers of this chapter to focus on the back-end development and the intelligence part of the system. The future direction will be focusing on improving the functionalities of the IPCS. Currently, it is able to match any selected items with all the available items in a store and produce a score/index for each pair. In real-life situations, co-ordinating several items to create a total look is common, thus an important research area is to explore the possibility of enhancing the ability of the inference engine to evaluate three or more items together.

Research can also be extended to incorporate the analysis of consumer buying behaviour and business intelligence based on the captured data. Fashion items, which have been tried on most or least, can be identified, and the data can be compared with the sales of these items for further analysis. The data obtained can be used to improve a number of analyses, including customer preferences and visual merchandising performance.

7.8 Acknowledgement

The work described in this chapter was fully supported by a grant from the Research Grants Council of the Hong Kong Special Administrative Region, China (Project No. PolyU 5226/08E).

7.9 Notes

[1] Cross-selling refers to selling of additional items to a customer in relation to the item(s) that the customer has purchased.

[2] Up-selling is a process through which a customer is persuaded (usually by a salesman) to purchase an upgrade of the item, which he/she intends to purchase.

[3] The middleware is used to allow seamless connections of various kinds of antennas and readers produced by different manufacturers.

7.10 References

Angeles, R. (2005), RFID technologies: Supply-chain applications and implementation issues, *Information Systems Management*, 22(1), 51–65.

Becker, J., Vilkov, L., Wei, B. and Winkelmann, A. (2010), A model based approach for calculating the process driven business value of RFID investments, *International Journal of Production Economics*, 127(2), 358–71.

Benbernou, R. and Warwick, K. (2007), A fuzzy multi-criteria decision approach for enhanced auto-tracking of seismic events, *IEEE International Conference on Signal Processing and Communications* (*ICSPC 2007*), pp. 1331–4.

Beynon, M., Cosker, D. and Marshall, D. (2001), An expert system for multi-criteria decision making using Dempster Shafer theory, *Expert Systems with Applications*, 20(4), 357–67.

Böse, I. and Mahapatra, R. (2001), Business data mining – a machine learning perspective, *Information & Management*, 39(3), 211–25.

Brown, I. and Russell, J. (2007), Radio frequency identification technology: An exploratory study on adoption in the South African retail sector, *International Journal of Information Management*, 27(4), 250–65.

Cakici, O. E., Groenevelt, H. and Seidmann, A. (2011), Using RFID for the management of pharmaceutical inventory – system optimization and shrinkage control, *Decision Support Systems*, doi: DOI: 10.1016/j.dss.2011.02.003.

Carlsson, C. and Fuller, R. (1995), On fuzzy screening system, *Proceedings of EUFIT'95 Conference*, Aachen: Verlag Mainz.

Chen, Y., Huang, T. and Chang, S. (2008), A novel approach for discovering retail knowledge with price information from transaction databases, *Expert Systems with Applications*, 34(4), 2350–9.

Cohen, M. (2004), Exploiting response models – optimizing cross-sell and up-sell opportunities in banking, *Information Systems*, 29(4), 327–41.

Gass, S. (2005), Model world: The great debate – MAUT versus AHP, *Interfaces*, 35, 308–12.

Guo, Z. X., Wong, W. K., Leung, S. Y. S. and Fan, J. T. (2009), Intelligent production control decision support system for flexible assembly lines, *Expert Systems with Applications*, 36(3), 4268–77.

Holmstrom, J., Framling, K. and Ala-Risku, T. (2010), The uses of tracking in operations management: Synthesis of a research program, *International Journal of Production Economics*, 126(2), 267–75.

Hui, S. C. and Jha, G. (2000), Data-mining for customer service support, *Information & Management*, 38(1), 1–13.

Jones, P., Clarke-Hill, C., Hillier, D. and Comfort, D. (2005), The benefits, challenges and impacts of radio frequency identification technology (RFID) for retailers in the UK, *Marketing Intelligence & Planning*, 23(4), 395–402.

Kamakura, W., Wedel, M., de Rosa, F. and Mazzon, J. (2003), Cross-selling through database marketing: A mixed data factor analyzer for data augmentation and prediction, *International Journal of Research in Marketing*, 20, 45–65.

Khandelwal, N., Krueger, K., Berger, B., Carpenter, M. and Butz, B. (2006), Developing hospice drug formulary using Multi-Attribute Utility Theory (MAUT) method and the evaluation of their economic impact, *Value in Health*, 9(6), A356–7.

Lee, I. and Lee, B.-C. (2010), An investment evaluation of supply chain RFID technologies: A normative modeling approach, *International Journal of Production Economics*, 125(2), 313–23.

Lefebvre, L. A., Lefebvre, É., Bendavid, Y., Wamba, S. F. and Boeck, H. (2007), The potential of RFID in warehousing activities in a retail industry supply chain, *Journal on Chain and Network Science*, 5(2), 101–10.

Lin, C. and Hong, C. (2008), Using customer knowledge in designing electronic catalog, *Expert Systems with Applications*, 34(1), 119–27.

Lin, Q., Chen, Y., Chen, J. and Chen, Y. (2003), Mining inter-organizational retailing knowledge for an alliance formed by competitive firms, *Information & Management*, 40(5), 431–42.

Loeb, W. F. (2003), Better customer service through technology, *Stores*, 85, 104.

Nasirin, S. and Birks, D. (2003), DSS implementation in the UK retail organisations: A GIS perspective, *Information & Management*, 40(4), 325–36.

Padmanabhan, B. and Tuzhilin, A. (1999), Unexpectedness as a measure of interestingness in knowledge discovery, *Decision Support Systems*, 27(3), 303–18.

Pei, J. and Klabjan, D. (2010), Inventory control in serial systems under radio frequency identification, *International Journal of Production Economics*, 123(1), 118–36.

Pictet, J. and Bollinger, D. (2008), Extended use of the cards procedure as a simple elicitation technique for MAVT: Application to public procurement in Switzerland, *European Journal of Operational Research*, 185(3), 1300–7.

RFID Journal (2002), Learning from Prada, available from: http://www.rfidjournal.com/article/view/425.

Rohacova, I. and Markova, Z. (2009), The analysis of the AHP method and its potential use in logistics, *ACTA Montanistica Slovaca*, 14(1), 103–12.

Sarac, A., Absi, N. and Dauzere-Peres, S. (2010), A literature review on the impact of RFID technologies on supply chain management, *International Journal of Production Economics*, 128(1), 77–95.

Seiter, L., Mezini, M. and Lieberherr, K. (2000), Dynamic component gluing, in: K. Czarnecki, K. Eisenecker and U. Eisenecker (eds), *Generative and Component-Based Software Engineering, First International Symposium, GCSE'99*, Berlin Heidelberg: Springer, pp. 134–46.

Simpson, L. (1996), Do decision-makers know what they prefer? MAVT and ELECTRE II, *Journal of the Operational Research Society*, 47(7), 919–29.

Tayma, J. F. and Pol, L. (1995), Retail site selection and geographic information systems, *Journal of Applied Business Research*, 11, 46–54.

Tong, R. (1982), Fuzzy-sets and multi-criteria decision-making, *Human Systems Management*, 3, 266–8.

Tsiporkova, E. and Boeva, V. (2006), Multi-step ranking of alternatives in a multi-criteria and multi-expert decision-making environment, *Information Sciences*, 176(18), 2673–97.

Wamba, S., Lefebvre, L., Bendavid, Y. and Lefebvre, E. (2008), Exploring the impact of RFID technology and the EPC network on mobile B2B eCommerce: A case study in the retail industry, *International Journal of Production Economics*, 112(2), 614–29.

Yin, S. Y. L., Tserng, H. P., Wang, J. C. and Tsai, S. C. (2009), Developing a precast production management system using RFID technology, *Automation in Construction*, 18(5), 677–91.

Zadeh, L. A. (1965), Fuzzy sets, *Information and Control*, 8, 338–53.

8

Measuring the impact of radio frequency identification (RFID) technologies in improving the efficiency of the textile supply chain

A. LUI and C. K. Y. LO, The Hong Kong Polytechnic University, Hong Kong

DOI: 10.1533/9780857098115.187

Abstract: The value of radio frequency identification (RFID) technology is critical for the clothing and textiles supply chain, because the fashion business is characterized by a wide assortment of seasonal items with short life-cycles, high levels of impulse-purchasing and complicated distribution and logistics operations. This study aims to test whether the adoption of the RFID system can improve supply chain performance (measured as inventory days, accounts receivable days and operating cycle). Based on the 31 clothing and textiles companies that publicly announced their RFID adoption, 18 of them were matched with comparable control firms for the testing of abnormal supply chain performance. The results show that clothing and textiles RFID adopters' inventory days drop by about 12.89 days over the 5-year period, while other RFID adopters (other manufacturing sectors) only drop by about 2.47 days.

Key words: RFID, event study, supply chain efficiency.

8.1 Introduction

Radio Frequency Identification (RFID) is an Auto-ID technology. It applies wireless radio frequencies to identify distant items from readers automatically. A RFID system usually consists of tags, readers and software for recording and analyzing information received from the readers. The tag is a microchip that is attached to a minuscule antenna. RFID was first introduced in 1948, but was not widely implemented in business operations until the 1990s (Landt, 2001). Today, RFID has become part of a commonly-adopted management system in sectors such as pharmaceuticals, logistics, fashion and retailing. Applications include assets management, payment, inventory management, customer services, production, etc. A study released by ABI Research (2010) predicted that spending on RFID products will grow at a compound annual growth rate of 15% between 2007 and 2013, becoming a US$9.7 billion industry by the end of 2013. Manufacturing sectors are rapidly adopting RFID, partly because of pressure from the major retailers such as Walmart and Marks & Spencer. These giant retailers have requested that their suppliers include RFID tags in their shipments to improve collaboration between retailer and suppliers in the supply chain.

The value of RFID technology is especially important in clothing and textiles manufacturing and retailing, because the fashion business is characterized by a wide assortment of highly seasonal products with short life-cycles, high levels of impulse-purchasing and complicated distribution and logistics operations (Moon and Ngai, 2008). In fashion and clothing retailing industries, item-based RFID implementation is an emerging trend. The RFID tagging of clothing by multinational clothing manufacturers and retailers such as American Apparel and Marks & Spencer is now in roll-out phase, with 200 million RFID labels used for clothing globally in 2009 (Das and Harrop, 2009). Clothing manufacturers are also under pressure to adopt RFID, due to mandates from giant retailers such as Walmart.

Except for a few suppliers, such as P&G, Unilever and Kimberly, which have declared benefits from adopting RFID, most suppliers still complain about high cost, unstable technology and limited benefits (Yang *et al.*, 2008). These complaints contradict the rapid growth of the RFID market and the benefits claimed by retail giants and RFID vendors. Some scholars argue that adopting RFID might only benefit retailers (downstream supply chain members) but not manufacturers (upstream supply chain members) (Whang, 2010), and that the rapid diffusion and market growth of RFID are due largely to institutional pressure (Barratt and Choi, 2007; DiMaggio and Powell, 1983; Lai *et al.*, 2006). Therefore, there is an urgent need to investigate and compare the true impact of RFID adoption on firm performance in different industries.

However, the impact of adopting RFID on operating performance has not yet been objectively estimated in written publications, and discussions on its impact remain unclear. Most previous studies focusing on RFID are either simply based on subjective judgments of its benefits from respondents or a few case studies focusing on a particular industry (Barratt and Choi, 2007). A thorough investigation of the longitudinal impact of RFID and comparative analysis between various industries have yet to be completed. There has only been one published RFID work based on objective financial data (Chang, 2011), with a sample size of only 65 manufacturing firms. Chang (2011) found adopting RFID improved firms' financial performance. However, because RFID technology itself is still unstable and the accuracy of reading the RFID tags subject to a number of constraints (i.e. the packaging, the angle of the reader toward the tags, and radio frequency absorptive material such as metals on the product packaging and its nearby environment), those financial benefits may only be due to improved sales performance after adoption.

The purpose of this study is to fill these gaps by addressing three research questions:

1. Does adopting RFID technology help firms improve their supply chain efficiency measured as inventory days, accounts receivable days and operating cycle time?
2. How does the impact of adopting RFID vary across the clothing industry (wholesalers and retailers vs. manufacturers)?

3. Do clothing and textile firms benefit most compared to other industries? The details of the measurement of supply chain efficiency will be discussed in the following section.

8.2 Methodology

8.2.1 Data collection

This study analyses all retailing and textile-related firms listed on the US stock market. The company list was generated from Standard and Poor's COMPUSTAT database. News articles on RFID investment announcements over the period from 1990 to 2012 were searched from periodical databases such as Factiva by using keywords such as RFID or radio frequency identification. Standard Industrial Classification (SIC) codes were used to classify the industries of the sample firms in this study. There were 520 companies in retailing industries and 83 of those companies were clothing retailers. We also included 97 companies from textile-related industries, making a total of 617 companies. Based on the standard practice of previous event studies (Corbett *et al.* 2005; Hendricks and Singhal, 2008), we focused on the first announcement among multiple announcements made by individual firms that adopted RFID. After screening we obtained 410 publicly listed firms that adopted RFID from 1990 to 2012.

8.2.2 Event study methodology

To investigate a firm's abnormal operating performance following its RFID investment announcements, this study uses event-study methodology recently employed in the IT and supply chain literature (Corbett *et al.*, 2005; Hendricks and Singhal, 2008). We followed the guidelines of Barber and Lyon (1996) for detecting abnormal operating performance. The event year (year 0) is defined as the year of RFID announcement (or successful implementation). To successfully implement RFID, the average preparation time is 6 to 12 months prior to complete operation. Therefore, year –2 is taken as the year before any RFID implementation. We are interested in comparing the operating performance before and after RFID implementation. As a result, the event period in this research is the time spanning the year of RFID implementation (year –2 to year 0) and the year of post-implementation (year 0 to year 3).

We matched sample and control pairs based on three matching criteria (pre-event performance, industry type and firm size) to minimize the effects of confounding factors in a particular industry or the effects of the overall state of the economy. Barber and Lyon (1996) suggested that matching pre-event performance is the most critical factor for event studies. They found that matching industry type and 90 to 110% pre-event performance created the most appropriate matching groups between sample and control firms. Therefore, we generated the sample-control pairs based on the following steps used by Barber and Lyon (1996). We first matched each sample firm to a portfolio of control firms based on at least a 2-digit SIC code, 90 to 110% of firm performance (e.g. inventory days) and 50 to 200% of the firm's total assets in the 2 years prior their adopting RFID. If no control firm was matched in Step 1, we relaxed the requirement of a 2-digit SIC code to one as our matching criteria. If no control firm was matched in Step 2, we used only 90 to 110% of performance and 50 to 200% of the firm's total assets as the matching criteria. Finally, if no comparable control firm existed after the above three steps, we chose the firm with the closest performance.

We calculated abnormal performance as the difference between sample post-event performance (i.e. actual performance in year 2) and expected performance (in year 2). We estimated expected performance as the sum of sample pre-event performance (i.e. in year –2) and change in the median performance of control firms during that period (i.e. from year –2 to year 3). The formulas are as follows:

$$AP_{(t+b)} = PS_{(t+b)} - EP_{(t+b)} \quad [8.1]$$

$$EP_{(t+b)} = PS_{(t+a)} + (PC_{k(t+b)} - PC_{k(t+a)}), \quad [8.2]$$

where:

- AP – abnormal performance;
- EP – expected performance;
- PS – performance of sample firms;
- PC – median performance of control firms of k number of control firms;
- t – year of RFID announcement;
- a – starting year of comparison ($i = -2, -1$ or 0);
- b – ending year of comparison ($j = -1, 0, 1, 2,$ or 3);
- k – number of control firms.

We present three statistical tests that are commonly used in event studies, namely the non-parametric Wilcoxon signed-rank (WSR) and sign test and the parametric t-statistics. The three statistics in the tables show similar results. We consider WSR as the most appropriate test for our study (Barber and Lyon, 1996), and the discussion below is based on the WSR statistics.

8.3 Industry distribution of RFID adoption in North America

Among the 410 RFID announcements that we collected from the Factiva and RFID journals, Fig. 8.1 presents the number of RFID announcements among the publicly listed manufacturers in North America. It shows that Food

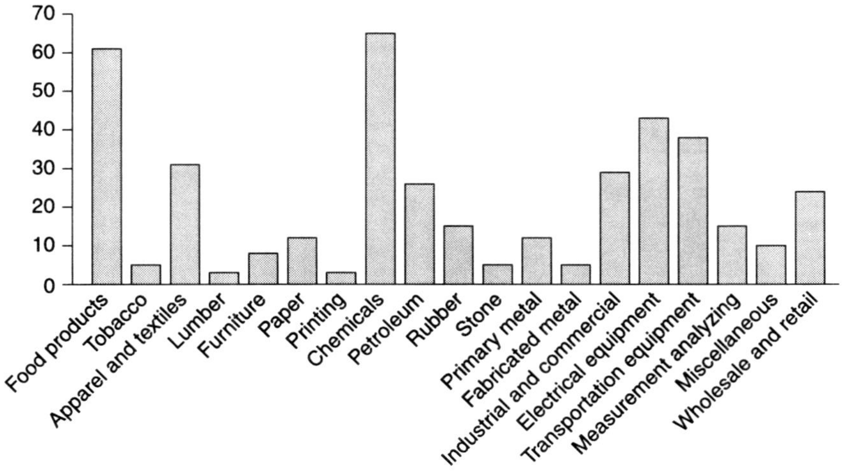

8.1 Distribution of 410 RFID announcements by industry.

Products (SIC code: 2000) and Chemicals (SIC: 2800) are the two industries with most RFID adoption announcements. Clothing and textiles-related industries (SIC: 2200, 2300, 3100, 5600) have 31 announcements. The distribution shows that although RFID technology could provide significant benefits to the fashion and textiles-related companies, it is still not commonly adopted in those industries. The figures need careful interpretation as the listed firms are mainly larger publicly listed firms, therefore the industry distribution of RFID adoption might not be the same for the small and medium enterprises (SMEs). Table 8.1 shows the specific number of announcements and the corresponding SIC code. Table 8.2 shows the year of distribution of the announcements and that the majority of the RFID adoption announcements came in the years 2004, 2005 and 2006. We believe that is mainly due to mandates from Walmart in 2003 requesting their top 200 suppliers to ship their products with RFID tags by the end of 2006.

8.3.1 Details of clothing and textiles firms which have adopted RFID

Table 8.3 shows all the announcements of RFID adoption in the clothing and textiles-related companies. The table contains the industry classification code (SIC code), the motivation of the adoption stated in the announcement, and whether they were Walmart suppliers who faced pressure from adoption mandates. Some firms became publicly listed after they adopted RFID, therefore no financial data is available for longitudinal analysis (that requires financial data for the company two years prior to their adoption of RFID); we have put a remark for

Table 8.1 Distribution by industry

Industry	SIC (2 digit)	Sample no.
Food products	2000	61
Tobacco	2100	5
Apparel and textiles	2200, 2300, 3100, 5600	31
Lumber	2400	3
Furniture	2500	8
Paper	2600	12
Printing	2700	3
Chemicals	2800	65
Petroleum	2900	26
Rubber	3000	15
Stone	3200	5
Primary metal	3300	12
Fabricated metal	3400	5
Industrial and commercial machinery	3500	29
Electrical equipment	3600	43
Transportation equipment	3700	38
Measurement analyzing	3800	15
Miscellaneous	3900	10
Wholesale and retail	5000–5900	24
Total		410

Table 8.2 Distribution by year

Year	Sample no.
1990	1
1992	1
1994	1
1996	2
1997	2
1998	7
1999	2
2000	7
2001	14
2002	8
2003	24
2004	65
2005	134
2006	63
2007	42
2008	20
2009	9
2010	5
2011	2
2012	1
Total	410

Table 8.3 List of apparel and textiles firms

No.	Company name	SIC	Adoption year	Motivation (1: Reative; 0: Proactive)	Mandator	Financial data (1: Available; 0: NA)
1	American Apparel Inc	2300	2008	0	None	0
2	Ashworth Inc	2300	2006	0	None	1
3	Bebe Stores Inc	2300	2004	1	NA	1
4	Calvin Klein Inc	2300	2005	0	None	0
5	Fruit of the Loom Ltd	2200	2005	1	NA	0
6	Garan Inc	2300	2005	1	NA	0
7	Jones Apparel Group Inc	2300	2009	1	Walmart	1
8	Kellwood CO	2300	2004	0	None	1
9	Levi Strauss & CO	2300	2004	1	Walmart	1
10	Liz Claiborne Inc	2300	2001	0	None	1
11	LVMH Moet Hennessy	2300	2001	0	None	1
12	Maidenform Brands Inc	2300	2012	0	None	0
13	Mohawk Industries Inc	2200	2005	0	None	1
14	Nautica Enterprises Inc	2300	2004	1	Walmart	1
15	Noel Group Inc	2200	2006	1	NA	0
16	North Face Inc	2300	2004	1	Walmart	0
17	Pillowtex Corp	2300	2005	1	NA	0
18	Playtex Apparel Inc	2300	2005	1	NA	0
19	Polo Ralph Lauren CP	2300	2006	0	None	1
20	Polymer Group Inc	2200	2005	0	None	1
21	Russell Corp	2200	2005	1	NA	1
22	Shaw Industries Inc	2200	2006	1	Walmart	0
23	Springs Industries	2200	2005	1	NA	0
24	Sunbeam Corporation	2300	2005	1	NA	0
25	Tandy Brands Accessories Inc	3100	2005	1	Walmart	1
26	Tommy Hilfiger Corp	2300	2008	0	None	0
27	VF Corp	2300	2005	1	Walmart	1
28	Wolverine World Wide	3100	2005	1	Department of Defense	1
29	J Crew Group Inc	5600	2005	0	None	1
30	Gap Inc	5651	2005	0	None	1
31	Abercrombie & Fitch	5651	2006	0	None	1

those records in the last column of Table 8.3. The data also shows that a number of famous fashion brands had already adopted RFID in their operations. There were 18 clothing and textile-related firms with financial data from year –2 to year 3 of RFID adoption.

8.4 Comparison of clothing and textiles RFID adopters with other RFID adopters

Table 8.4 shows the characteristics between clothing and textiles firms adopting RFID and other firms adopting RFID. It tries to address the issue of whether more productive or profitable firms tend to adopt the technology. We compared firm characteristics, such as size, profitability, R&D intensity, etc. of the clothing and textiles RFID adopters to other RFID adopters. We found that clothing and textiles RFID adopters are significantly smaller than other firms adopting RFID. The labour productivity and R&D intensity of clothing and textiles RFID adopters are also significantly lower than other RFID adopters.

We also investigated the financial performance between clothing and textiles RFID adopters and other RFID adopters. We conducted t-tests on the return on assets (ROA), return on sales (ROS), sales over assets (SOA) and sales growth between the two sets of adopters. We only found a difference in SOA between clothing and textiles RFID adopters and other RFID adopters. Clothing and textiles RFID adopters show significantly higher SOA than other RFID adopters, which means that clothing and textiles RFID adopters can generate more revenue with the same amount of firm assets compared to other RFID adopters. However, the overall profitability (ROA) and profit margin (ROS) show no difference. Therefore, clothing and textiles RFID adopters have great potential to improve profit margin and productivity, as they are able to mark higher prices for their products compared to other industries.

To further investigate the missing link between the higher SOA and lack of difference in ROA, we looked into the supply chain efficiency indicators. We compared inventory days, accounts receivable days and operating cycle between clothing and textiles RFID adopters and other RFID adopters. Inventory days and accounts receivable days are calculated by dividing 365 days by the inventory turnover ratio (or accounts receivable turnover) from Standard and Poor's COMPUSTAT database. The results show that clothing and textiles RFID adopters' median inventory days is 96.52 days, while other RFID adopters are just 58.03 days in year –2, a difference of nearly 40 days, indicating that clothing and textiles RFID adopters have huge room for improvement in supply chain efficiency. We also compared the accounts receivables days, but saw no significant difference. Finally, the operating cycle time of clothing and textiles RFID adopters is 144.73 days, while other RFID adopters is 110.62 days, which shows that the major difference is in the inventory days. This finding suggests that if RFID can improve clothing and textiles-related firm performance, the most likely indicator would be an improvement in inventory days.

Table 8.4 Difference between apparel and textiles RFID adopters and other RFID adopters in year –2

	N	Mean	Median	*p*-Value (t-test)	*p*-Value (Mann-Whitney test)	*p*-Value (Kolmogorov-Smirnov test)
Firm size						
Apparel and textiles firms	18	3,424	1,765	0.000***	0.001***	0.008***
Other RFID adopters	245	20,205	7,628			
Labour productivity						
Apparel and textiles firms	18	24.78	19.75	0.000***	0.001***	0.024**
Other RFID adopters	215	68.37	39.90			
R&D intensity						
Apparel and textiles firms	18	0.10	0.00	0.456	0.000***	0.000***
Other RFID adopters	199	8.15	1.94			
ROA						
Apparel and textiles firms	18	17.96	17.58	0.375	0.102	0.235
Other RFID adopters	225	13.06	13.45			
ROS						
Apparel and textiles firms	18	12.92	12.38	0.789	0.565	0.557
Other RFID adopters	225	13.49	13.60			
SOA						
Apparel and textiles firms	18	144.34	144.83	0.242	0.011**	0.005***
Other RFID adopters	244	123.65	102.26			
Sales growth						
Apparel and textiles firms	18	9.63	9.71	0.742	0.676	0.912
Other RFID adopters	239	8.39	7.91			
Inventory days						
Apparel and textiles firms	18	119.25	96.52	0.001***	0.001***	0.004***
Other RFID adopters	238	73.01	58.03			

(*Continued*)

Table 8.4 Continued

	N	Mean	Median	*p*-Value (t-test)	*p*-Value (Mann-Whitney test)	*p*-Value (Kolmogorov-Smirnov test)
Accounts receivable days						
Apparel and textiles firms	17	38.75	42.54	0.116	0.149	0.207
Other RFID adopters	236	51.98	47.55			
Operating cycle						
Apparel and textiles firms	17	125.14	144.73	0.045**	0.033**	0.045**
Other RFID adopters	235	161.07	110.62			

*$p<0.1$; **$p<0.05$; ***$p<0.01$

8.5 Supply chain performance differences between apparel and textiles RFID adopters and general manufacturing RFID adopters

To test whether inventory days, accounts receivable days and the operating cycle of the clothing and textiles-related RFID adopters actually improved after adopting RFID, we conducted a long-term event study analysis of the firms' supply chain efficiency indicators. We first analysed the impact on supply chain performance (e.g. inventory days) of clothing and textiles RFID adopters over a 5-year period (i.e. 2 years prior to RFID adoption and 3 years after RFID adoption). We then compared clothing and textiles RFID adopters and other RFID adopters in general. Table 8.5 presents the results of the cumulative abnormal changes in inventory days, accounts receivable days and operating cycle.

The data clearly shows that both clothing and textiles RFID adopters and other RFID adopters can improve the inventory days over the 5-year period. Clothing and textiles RFID adopters' inventory days drop by about 12.89 days over the 5-year period, while other RFID adopters only drop by about 2.47 days. Although both groups show significant changes, the clothing and textiles RFID group has clearly shown much stronger changes over the period. The abnormal changes in accounts receivable days and operating cycle time of clothing and textiles RFID adopters are also significant in some tests, but the changes are less obvious compared to the changes in inventory days. The difference in these two indicators between clothing and textiles RFID and other RFID adopters also show that the clothing and textiles RFID adopter group experienced a much greater improvement in their operations. Tables 8.6 and 8.7 give the results of the whole sample and manufacturing firms, respectively, which consistently show firms experience improvement on supply chain performance.

Finally, we compared the changes between wholesaler and retailer samples with the manufacturing sample. As shown in Table 8.8, although the wholesalers and retailers is a much smaller sample size, the abnormal changes in inventory days and operating cycle time are clearly statistically significant. In contrast, manufacturing firms only experienced significant changes in inventory days but not the operating cycle. This finding corroborated the discussion in the RFID literature that retailers reap more benefits than manufacturers (Whang, 2010).

8.6 Conclusion

The present long-term event study is the first attempt to provide a thorough review and quantitative evidence examining and comparing the impact of RFID adoption on supply chain efficiency across the fashion, retail and

Table 8.5 Cumulative abnormal changes in inventory days, account receivable days and operating cycle

Apparel and textiles firms							Other RFID adopters					
Time period	N	Abnormal mean	Abnormal median	*p*-Value (t-test)	*p*-Value (WSR test)	*p*-Value (sign test)	N	Abnormal mean	Abnormal median	*p*-Value (t-test)	*p*-Value (WSR test)	*p*-Value (sign test)
Inventory days												
t−2 to *t*+1	16	−8.62	−11.74	0.073*	0.063*	0.210	213	−3.92	−1.01	0.096*	0.171	0.040
t−2 to *t*+2	10	−10.00	−7.12	0.059*	0.059*	0.344	196	−7.18	−2.55	0.011**	0.010**	0.018**
t−2 to *t*+3	12	−13.75	−12.89	0.048**	0.084*	0.388	178	−6.10	−2.47	0.044**	0.005***	0.001***
Accounts receivable days												
t−2 to *t*+1	14	−11.47	−2.82	0.151	0.056*	0.424	197	−2.49	−1.26	0.053*	0.142	0.476
t−2 to *t*+2	11	−5.36	−4.50	0.022**	0.033**	0.227	183	−2.04	−0.29	0.189	0.443	0.657
t−2 to *t*+3	9	−5.19	−3.62	0.118	0.066*	0.180	148	−3.94	−1.66	0.061*	0.065*	0.162
Operating cycle												
t−2 to *t*+1	14	−9.62	−8.81	0.007***	0.011**	0.013**	207	−1.53	−0.59	0.615	0.810	0.578
t−2 to *t*+2	10	−5.15	−7.38	0.478	0.386	0.344	186	−6.40	−1.67	0.065*	0.204	0.420
t−2 to *t*+3	9	−16.2	−14.48	0.076*	0.066*	0.180	168	−5.90	−3.65	0.133	0.093*	0.247

*$p<0.1$, **$p<0.05$, ***$p<0.01$

Table 8.6 Whole sample (apparel and textile, wholesalers and retailer and manufacturing firms)

Time period	N	Abnormal mean	Abnormal median	*p*-Value (t-test)	*p*-Value (WSR test)	*p*-Value (sign test)
			Inventory Days			
t−2 to *t*+1	229	−4.24	−1.28	0.056*	0.071*	0.017**
t−2 to *t*+2	206	−6.15	−2.64	0.010**	0.007***	0.015**
t−2 to *t*+3	190	−7.53	−2.92	0.006***	0.001***	0.000***
			Accounts Receivable Days			
t−2 to *t*+1	211	−3.09	−0.86	0.018**	0.058*	0.271
t−2 to *t*+2	194	−2.25	−0.61	0.126	0.223	0.351
t−2 to *t*+3	157	−4.05	−2.02	0.042**	0.031**	0.055*
			Operating Cycle			
t−2 to *t*+1	221	−2.04	−1.01	0.478	0.453	0.226
t−2 to *t*+2	196	−6.34	−2.50	0.055*	0.151	0.284
t−2 to *t*+3	177	−6.44	−4.76	0.086*	0.042**	0.133

*$p<0.1$, **$p<0.05$, ***$p<0.01$

Table 8.7 Manufacturing sample firms (include Apparel and Textiles)

Panel A Abnormal Changes in Inventory Days

Time period	N	Abnormal mean	Abnormal median	*p*-Value (t-test)	*p*-Value (WSR test)	*p*-Value (sign test)
			Yearly abnormal change			
t−3 to *t*−2	238	0.11	0.98	0.944	0.708	0.270
t−2 to *t*−1	232	0.66	0.45	0.531	0.972	0.555
t−1 to *t*	225	0.09	0.27	0. 943	0.537	0.689
t to *t*+1	207	−5.00	−0.74	0.000***	0.009***	0.095*
t+1 to *t*+2	190	−1.69	−0.56	0.152	0.323	0.717
t+2 to *t*+3	174	−3.047	−2.28	0.039**	0.005***	0.005***
			Cumulative abnormal change			
t−2 to *t*+1	207	−1.01	−5.02	0.034**	0.103	0.071*
t−2 to *t*+2	190	−7.17	−2.50	0.015**	0.023**	0.050**
t−2 to *t*+3	174	−6.74	−2.47	0.033**	0.004***	0.002***

Panel B Abnormal Changes in Accounts Receivable Days

Time period	N	Abnormal mean	Abnormal median	*p*-Value (t-test)	*p*-Value (WSR test)	*p*-Value (sign test)
			Yearly abnormal change			
t−3 to *t*−2	232	0.76	0.57	0.216	0.518	0.393
t−2 to *t*−1	224	0.37	−0.23	0.505	0.775	0.738
t−1 to *t*	215	−0.43	0.53	0.578	0.445	0.413
t to *t*+1	196	−3.37	−0.50	0.001***	0.005***	0.054*
t+1 to *t*+2	178	−1.27	−0.48	0.120	0.311	0.500
t+2 to *t*+3	139	−1.70	−1.67	0.014**	0.012**	0.090*

(*Continued*)

Table 8.7 Continued

Time period	N	Abnormal mean	Abnormal median	*p*-Value (t-test)	*p*-Value (WSR test)	*p*-Value (sign test)
			Cumulative abnormal change			
$t-2$ to $t+1$	196	−3.17	−1.04	0.024**	0.080*	0.353
$t-2$ to $t+2$	178	−1.35	−0.84	0.306	0.361	0.500
$t-2$ to $t+3$	139	−2.42	−2.02	0.034**	0.033**	0.062*

Panel C Abnormal Changes in Operating Cycle

Time period	N	Abnormal mean	Abnormal median	*p*-Value (t-test)	*p*-Value (WSR test)	*p*-Value (sign test)
			Yearly abnormal change			
$t-3$ to $t-2$	234	1.68	1.47	0.296	0.324	0.170
$t-2$ to $t-1$	231	1.63	0.83	0.290	0.540	0.792
$t-1$ to t	224	1.10	1.51	0.510	0.241	0.180
t to $t+1$	202	−6.41	−2.29	0.001***	0.006***	0.057**
$t+1$ to $t+2$	180	−5.16	−1.99	0.009***	0.090*	0.333
$t+2$ to $t+3$	160	−5.02	−2.95	0.036**	0.017**	0.033**
			Cumulative abnormal change			
$t-2$ to $t+1$	202	−0.27	−0.42	0.903	0.849	0.778
$t-2$ to $t+2$	180	−4.45	−0.72	0.161	0.399	0.709
$t-2$ to $t+3$	160	−7.79	−4.81	0.039**	0.087*	0.304

*$p<0.1$, **$p<0.05$, ***$p<0.01$

manufacturing industries. According to the data collected from 410 firms adopting RFID in the US from 1990 and 2012, RFID adoption clearly has a positive impact on the supply chain efficiency in the long term, and clothing and textiles firms in particular can benefit more in their operations than other adopters in general. The results suggest that RFID technology is a key factor to enhance competitive advantage in supply chain performance for clothing and textiles firms. These results are also important to managers in the clothing and textiles industry, as they face increasing pressures from supply chain partners to adopt RFID.

Table 8.8 Subsamples of other RFID adopters

Wholesalers and retailers							Manufacturing firms					
Time period	N	Abnormal mean	Abnormal median	*p*-Value (t-test)	*p*-Value (WSR test)	*p*-Value (sign test)	N	Abnormal mean	Abnormal median	*p*-Value (t-test)	*p*-Value (WSR test)	*p*-Value (sign test)
Inventory days												
$t-2$ to $t+1$	17	−5.42	−4.77	0.009***	0.006***	0.002***	196	−4.44	−0.90	0.134	0.304	0.175
$t-2$ to $t+2$	15	−4.90	−6.30	0.005***	0.006***	0.035**	181	−8.91	−2.49	0.024**	0.030**	0.074*
$t-2$ to $t+3$	14	−5.63	−6.24	0.004***	0.009***	0.013**	164	−9.35	−2.26	0.028**	0.006***	0.002***
Accounts receivable days												
$t-2$ to $t+1$	14	−1.64	−1.00	0.132	0.140	0.180	183	−2.37	−0.60	0.082*	0.211	0.657
$t-2$ to $t+2$	14	−2.35	−0.99	0.130	0.245	0.424	169	−1.91	−0.09	0.250	0.588	0.878
$t-2$ to $t+3$	11	−5.16	−6.87	0.053*	0.062*	0.227	137	−3.82	−1.13	0.090*	0.136	0.305
Operating cycle												
$t-2$ to $t+1$	15	−6.61	−7.04	0.014**	0.015**	0.035**	192	−1.03	0.14	0.754	0.814	0.942
$t-2$ to $t+2$	14	−8.40	−6.13	0.008***	0.008***	0.057*	172	−6.10	−0.71	0.103	0.439	0.819
$t-2$ to $t+3$	14	−10.29	−8.60	0.005***	0.008***	0.057*	152	−4.99	−2.28	0.126	0.223	0.570

*$p<0.1$; **$p<0.05$; ***$p<0.01$

8.7 References

ABI Research.com. (2010), *RFID Market to Reach $5.35 Billion this Year*, vol. 2011, New York: ABI Research.

Barber, B. M. and Lyon, J. D. (1996), Detecting abnormal operating performance: The empirical power and specification of test statistics, *Journal of Financial Economics*, 41(3), 359–99.

Barratt, M. and Choi, T. (2007), Mandated RFID and institutional responses: Cases of decentralized business units, *Production and Operation Management*, 16(5), 569–85.

Chang, Y. (2011), Does RFID improve firms' financial performance? An empirical analysis, *Information Technology and Management*, 12(3), 273–85.

Corbett, C. J., Montes-Sancho, M. J. and Kirsch, D. A. (2005), The financial impact of ISO 9000 certification: An empirical analysis, *Management Science*, 51(7), 1046–59.

Das, R. and Harrop, P. (2009), *RFID Forecasts, Players and Opportunities 2009–2019*, vol. 2011: IDTechEX.

DiMaggio, P. J. and Powell, W. W. (1983), The iron cage revisited: Institutional isomorphism and collective rationality in organizational fields, *American Sociological Review*, 48(2), 147–60.

Hendricks, K. B. and Singhal, V. R. (2008), The effect of product introduction delays on operating performance, *Management Science*, 54(5), 878–92.

Lai, K., Wong, C. W. Y. and Cheng, T. C. E. (2006), Institutional isomorphism and the adoption of information technology for supply chain management, *Computers in Industry*, 57(1), 93–8.

Landt, J. (2001), *Shrouds of Time: The History of RFID*, An AIM Publication.

Moon, K. L. and Ngai, W. T. (2008), The adoption of RFID in fashion retailing: A business value-added framework, *Industrial Management & Data Systems*, 108(5), 596–612.

Whang, S. (2010), Timing of RFID adoption in a supply chain, *Management Science*, 56(2), 343–55.

Yang, X., Kishore, R. and Liu, Z. (2008), *The Institutional Facets of Innovation Diffusion Initiating: The Case of Wal-Mart's RFID Campaign*, Paper presented at the ICIS 2008 Proceedings.

9

Modeling the effectiveness of radio frequency identification (RFID) technologies in improving sales performance in fashion retail outlets

A. DE MARCO, A. C. CAGLIANO, M. L. NERVO and C. RAFELE, Politecnico di Torino, Italy

DOI: 10.1533/9780857098115.203

Abstract: The application of item-level radio frequency identification (RFID) technology in retail supply chains is beneficial in terms of cost savings and increased revenue. However, the economic effects of improved store operations, labor utilization and increased sales have yet to be fully explored. This chapter uses system dynamics (SD) as a means of carrying out structural modeling and simulation, to be integrated with conventional evaluations of return on investment (ROI). Building on previous research into RFID technology in retail supply chains, a model is developed on the basis of a case study of a leading Italian apparel retailer. Simulations show that RFID implementations are profitable whenever they contribute to increasing sales, especially when a fashion retailer is focused on clerk-assisted sales strategies. RFID technology allows better inventory control and faster inventory turnover; moreover, as a result of these more efficient backroom operations, store personnel have more time to assist consumers, leading to a growth in sales.

Key words: radio frequency identification (RFID), supply chain management, retail operations, fashion industry, system dynamics (SD).

9.1 Introduction

RFID technology has the potential to make a significant impact on retail supply chain (SC) operations.[1] By remotely reading the electronic product code and other pieces of information on a tag and making data available to information systems across the SC, RFID technology claims to allow better inventory control, reduce stock-out, offer savings in labor costs, and to lead to reduced shrinkage and fewer transaction errors. However, despite several pilot studies and deployment in the retail sector, there is still a limited understanding of the business value of RFID technology (Delen *et al.*, 2007; Lee and Özer, 2007; Whitaker *et al.*, 2007) and, in particular, of item-level RFID tagging. While the benefits that SCs gain from automation of operational processes and from better information are quantifiable

[1] Note: This chapter is adapted from: A. De Marco, A. C. Cagliano, M. L. Nervo and C. Rafele (2012), Using system dynamics to assess the impact of RFID technology on retail operations, *International Journal of Production Economics*, 135(1), 333–44. Reproduced with permission from Elsevier.

(Visich *et al.*, 2009), item-level RFID implementation at the shop floor level requires further investigation. In particular, RFID tagging at the item level promises to bring significant benefits to retailers in enhancing shop floor productivity, promoting sales and improving business performance in general (Gaukler *et al.*, 2007).

Numerous studies have been carried out that explore cost-benefit analysis (Özelkan and Galambosi, 2008), the process-driven value of RFID investments across SC stakeholders (Becker *et al.*, 2010), and the optimal configuration of RFID systems within a supply chain network (Chang *et al.*, 2010). However, there is little research available that evaluates the economic impact of item-level RFID-enabled retail store operations on sales and revenue performance. There are two principal reasons for this. First, the contribution of RFID to sales growth is influenced by a number of interrelated factors, making it difficult to identify the separate benefits offered by individual RFID functions. It is therefore more useful to consider retail store operations as a complex system demanding complex analysis techniques. Second, sales performance is heavily dependent on specific SC organizations and retail strategies. Evaluation frameworks and general business cases should thus be built into company-based case studies (Tzeng *et al.*, 2008).

To overcome these issues, this chapter first proposes the use of system dynamics (SD) (Forrester, 1961; Sterman, 2000) as a modeling and simulation tool to refine *ex-ante* estimates. SD is the preferred approach, because conventional financial methods, such as capital budgeting with discounted cash flow analysis (CFA), do not inherently include any feedback relationships between system variables in their analysis. In contrast, SD primarily aims to investigate the structure of these cause-and-effect relations and how they evolve over time. Specifically, SD is used here to understand the extent to which RFID technology might impact on the complex relationships between retail operations, inventory management and sales performance, and to provide accurate input for CFA.

We then build a model based on a case study of implementing item-level RFID technology into the retail stores of a leading Italian Apparel and Fashion (A&F) retailer: Miroglio Fashion S.r.l. This is part of a larger project to restructure the supply chain that the company started in 2007. The project offers a valuable opportunity to illustrate the integrated effects of RFID on sales performance, inventory management and sales force utilization. We also show how the new technology might affect retail business growth, depending on the fashion sales strategy adopted by a particular retailer. The case study serves to prove that SD is a practical and predictable means of assisting in the evaluation of the cost and revenue benefits that can be achieved through the use of RFID technology in retail stores.

The case study is presented as follows. Section 9.2 reviews relevant research in the field and looks at the value of RFID technology to date. Section 9.3 outlines the research methodology adopted in this chapter for the modeling of retail store operations. Section 9.4 introduces the case study, while Section 9.5 discusses the model for the project, and Section 9.6 the procedures for testing the model. The next two sections discuss the findings, with Section 9.7 focusing on the impact of

RFID, while Section 9.8 looks specifically at the question of return on investment (ROI). Finally, the conclusions are summarized in Section 9.9, along with some limitations and potential future directions.

9.2 Assessing the value of RFID technology

Within the vast amount of literature on the adoption and application of RFID technology (Curtin *et al.*, 2007; Ngai *et al.*, 2008), studies specifically related to the problem of assessing the economic impact of RFID technology can be subsumed into two main management areas: those concerned with understanding the roots of RFID-enabled cost efficiency in retail SCs, and those that look at the impact of RFID on retail store operations and the revenue stream.

The first group can in turn be broken down into further streams of research. Some studies explore the strategic and operational value of RFID to reduce cost and improve the service performance of inter-organizational SCs (Ju, 2008) through enhanced information sharing, increased asset utilization and higher visibility (Lee, 2009; Tzeng *et al.*, 2008; Ustundag, 2009). A second stream of research is more focused on the use of quantitative methods to measure the economic benefits resulting from different aspects of RFID implementation (Delen *et al.*, 2007; Doerr *et al.*, 2006; Kumar *et al.*, 2011; Lee, 2009; Prater and Frazier, 2005) and to provide ROI evaluation models (Lee and Lee, 2010). Within this stream, some works are specifically related to fast retail (Bottani and Rizzi, 2008) and fashion businesses (Koh *et al.*, 2006; Moon and Ngai, 2008). In particular, Bottani *et al.* (2009) examine a fashion SC composed of a distribution center and a retail store, in order to assess the profitability of an RFID investment and to provide guidelines for its implementation in the A&F industry.

Fieldwork initiatives and pilot studies have also been carried out to determine the effectiveness of RFID implementation and the achievement of cost savings. Key examples come from the experiences of Wal-Mart and METRO Group (Hardgrave *et al.*, 2008; Ton *et al.*, 2005). Some cases report on pilots in the A&F retail industry, such as Prada, Benetton, Marks & Spencer and Kaufhof (Loebbecke and Huyskens, 2006; Loebbecke, 2008). With specific regard to Galeria Kaufhof, Thiesse *et al.* (2009) survey the use of RFID at the intersection between store logistics and customer service, analyze physical in-store processes, put forward theories about the effects that RFID may have on business processes, and develop a conceptual model to explain the different cause-and-effect chains between RFID investments and their impact on business performance. Gaukler *et al.* (2007) provide an extensive list of reports and academic papers on RFID trials in the retail industry.

Turning to the second major research area, most models and case-based studies demonstrate that RFID technology has an impact on profitability through revenue growth rather than through cost reduction (Mithas *et al.*, 2008; Tzeng *et al.*, 2008). It is therefore important to assess the impact of RFID on sales and revenues (Whitaker *et al.*, 2007). This second area therefore looks at the benefits that RFID can offer in retail store operations and the revenue stream. It seems that item-level RFID technology contributes to revenue growth because of its inherent intelligent ability to improve sales and capture the customer's needs (Lo *et al.*, 2008). For example, Koschat (2008) demonstrates that a greater inventory leads to increased consumer demand. In particular, A&F consumers perceive large quantities of items displayed on shelves as an indication of novelty and a major motivation for shopping. Item-level RFID tags can therefore contribute to sales growth through improvements in inventory management and on-shelf availability. For instance, some business studies carried out on pilot schemes (Bauer *et al.*, 2008) report that revenues can increase by 5 to 10% when RFID technology is used to improve store inventories.

However, claims that RFID investments contribute not just to cost savings but also to revenue expansion have not yet been completely substantiated: detailed model-based analysis is required, which would allow integration of the CFA with the complex relationships between different areas of the business influenced by RFID technology (Lee and Özer, 2007). This chapter seeks to fill this gap by using a case-based model to understand the extent to which item-level RFID technology provides value through sales growth and cost efficiency in fast fashion trades (Rust *et al.*, 2002). The case study allows a model to be built to simulate the complex relationships between the variables that affect business performance and to provide CFA with pertinent input data.

9.3 Modeling retail store operations

The company examined in the case study undertook a large project late in 2007, with the aim of restructuring the SC through the implementation of item-level RFID technologies. As part of this project, our research group was given the task of investigating the strategic impact of RFID technology on store operations and providing a viable method to assess the benefits for both revenue and cost streams.

Our approach was based on SD model and simulation phased methodology, as described by Lyneis (1998). SD is a modeling and computer-based simulation method that helps in the understanding of complex systems (Sterman, 2000). SD permits the graphic diagramming of a system of feedback causal loops between interrelated accumulation stocks, flow rates, and auxiliary variables, and the definition of various linear and nonlinear mathematical relations. It also allows commercial software packages to carry out the discrete-step computation required to solve differential sets of equations over a preset time frame (Sterman, 2000). The output from the computer simulation provides a graph showing the curve lines of all variables plotted on the time axis, and the model testing is based on historical data and sensitivity analyses. SD lets the user understand the overall dynamics of the system and the impact of the different variables on the problem at issue; it also provides support for important decisions, and offers a means of testing policies by simulating different scenarios. SD has been widely used in production and operations management to explore inventory capacity and instability (Anderson *et al.*, 2005; Croson and Donohue, 2005; White 1999), as well as to investigate the effects of new technologies on business strategies (Disney *et al.*, 2004; Lyneis, 1998; Pardue *et al.*, 1999; Winch, 1997).

Following the approach adopted by Lyneis, we first worked on understanding, structuring and analyzing the system with regard to distribution, store operations and sales issues. This was achieved by gathering detailed information and process flow data, process mapping, interviews with logistics managers, marketing executives and salespersons, as well as through direct observation of warehouse, shipping and retail store operations.

A first-hand conceptual causal loop diagram was developed with the participation of the company's personnel, and a numerical SD model was then worked out. In compliance with SD principles and practice, the case model included stock variables, flows and feedback loops that tackle store inventory management, store labor utilization and customer demand, and is based on a select number of previous studies. In part, it is founded on the 'Stock Management Structure' model devised by Sterman (2000), which replicates a supply line integrated with stock control balancing feedback loops, comprising both the desired and actual adjustment quantities and the review periods examined. However, the model follows standard practice in SD, rather than other approaches, in considering customer demand as an endogenous variable affected by the performance of the supply line itself (Akkermans and Dellaert, 2005; Ashayeri and Lemmes, 2006; Gonçalves *et al.*, 2005). For instance, Gonçalves *et al.* (2005) demonstrate that although SC instability and customer demand are often analyzed separately, both interact with product availability: product shortages cause lower demand, which in turn leads to reduced production, prolonging stock-outs that further lower demand, as a destabilizing reinforcing feedback loop, resulting in demand being endogenously affected from within the SC system.

The mathematical equations that underpin the stock and flow diagram were then drawn up. The model was then tested by a variety of methods: curve fitting of historical data, robustness appraisal in relation to extreme values, and analysis of model sensitivity associated with random exogenous variables. The next step was to carry out a number of simulation runs using different scenarios to investigate the effects of RFID implementation on inventory management, use of labor and sales growth. The simulations also provided input data for CFA and ROI assessments. The results were then analyzed and recommendations made.

9.4 Case study: Miroglio Fast Fashion

The mass A&F market is a buyer-driven SC, where retailers seek to capture the consumers' mood of the moment. Products in the market typically have short life-cycles with highly volatile and seasonal demand; impulse purchasing is also a key feature. Mass apparel retailers therefore follow fashions and seek to exploit the market by bringing new products to their stores as frequently as possible. A large variety of clothes must be designed at short notice so as to include the most recent fashion trends. They must also be made available as quickly as possible to meet customers' requirements, and in sufficient quantities to ensure sufficient sales, avoid stock-outs and allow prompt replenishment of successful items (Christopher *et al.*, 2004). These factors have all led to an increasing focus on the agile SC and the philosophy of the quick response, a means of increasing the speed and flexibility of reactions to market shifts (Lowson *et al.*, 1999; Masson *et al.*, 2007).

Key international competitors in the mass fast fashion market include Zara-Inditex, H&M-Hennes & Mauritz, The Gap, Benetton, Miroglio and Mango. Most follow an innovative business plan featuring in-house creative design, centralized distribution and frequent shipments to well-located stores that are attractive to consumers (Ghemawat and Nueno, 2006). The stores themselves in fact play a crucial role in communication and advertising: they aim to offer customers a unique shopping experience through variety and novelty of products, to stimulate brand loyalty and to foster the retailer's identity and brand equity compared to their competitors. The mass fast fashion market displays a number of features, including demand volatility, flexibility and speed, which make it an ideal testing ground for the validity of SD techniques in assessing the effectiveness of RFID technology.

Miroglio Fashion S.r.l., Fast Fashion Division (Miroglio) is a global retailer headquartered in Alba, Italy, which designs, manufactures and sells women's garments and accessories through the Motivi, Oltre and Fiorella Rubino brands. At the end of 2008, it produced 20 million clothing items and operated more than 1680 stores globally (although this study is restricted to only the 850 stores located in Italy). In the two seasons, fall/winter and spring/summer, each brand chain presents several collections, which are in turn split into a variety of themes, or looks, that differ in design, fabrics and colors. A centralized distribution center (DC) located close to the headquarters collects from multiple manufacturers and packs items to serve all retail stores around the world (Cagliano, 2010).

Product novelty, along with frequency of consumer visits, is achieved by shipping a new theme to Miroglio stores every two weeks, and reconfiguring the interior display. In addition, store managers and sales assistants are committed to delivering experienced and personalized advice to customers, in order to enhance brand fidelity. The corporate managers of the business claim that this clerk-assisted sales strategy distinguishes Miroglio stores from those of its competitors, which are more based on self-purchasing and visual merchandising (Mittelstaedt and Stassen, 1990), and has the added advantage of involving salespersons in the process of improving product quality. The shop personnel must therefore be dedicated to assisting customers, and not required to spend very much time on backroom tasks and other operations that are not directly aimed at selling products.

We now turn to a description of inventory control and receiving and other operations as they are currently performed in Miroglio's stores. With regard to inventory management, the initial quantity and variety of items to fill the store inventory at the beginning of each theme period (9 per season) are determined centrally by the marketing department, based on demand planning. Replenishment of best-selling items can then be carried out up to twice a day. Store managers are responsible for determining when a replenishment order is required, based on personal experience, past sales performance and forecast, as well as projected store window displays. The reorder quantity is checked and may be adjusted by the central marketing, the basis of store location and visitor frequency statistics. Orders enter the DC queuing system that optimizes the number and size of shipments based on the tradeoff between time and cost of transportation. Whenever a product goes out of stock, availability at other nearby stores is checked; an urgent order can then be placed, which will be sent out in the shortest possible timeframe.

At present, unit loads received in the store backroom are checked against the bill of lading. Then, whenever sales staff are available, items are removed from boxes, individual barcodes are scanned and magnetic-acoustic security tags are applied. A portion of the items received is soon displayed on shelves in the store, while the rest is stored in the backroom. Typically, 70% of the total store inventory is displayed on the shop floor. All salespersons share backroom duties to facilitate the selection of items based on customer demand. At check out, items are scanned for payment transaction and the security tags are removed. Unsold items are put on promotional sale for an 8-week period at the end of the season; any remaining items, which make up approximately 10% of total production, are returned: security tags are removed and a barcode-scanned list of lading is created. Table 9.1 provides a summary of the main tasks with associated parameters, as recorded by the company's internal auditors of both the Sales and Marketing and the Logistics departments, over a 4-year period since 2005.

We now turn to the impact of introducing RFID technology into the Miroglio SC, with specific focus on assessing the benefits that can result from more efficient shop operations and, in particular, on understanding the sources of sales growth. A disposable garment-care RFID tag was selected (passive ultra-high frequency EPC global standard UHF Class2 Gen2) for this purpose. Handheld readers, with a reading distance of up to 2 meters and a capacity to scan 250 items simultaneously, are used to support the store staff in the receipt of goods as well as with treatment, picking, inventory taking, item positioning and dressing of shop windows. Information corners and theft prevention portals are also included.

With this configuration, a few aspects of store operations will undergo significant changes: incoming deliveries will not need to be counted manually, meaning that boxes do not have to be opened, and items will be made available for display more quickly, because price labeling will be replaced by information corners. In addition, anti-theft magnetic tagging will no longer be necessary, since the RFID tag incorporates theft prevention technology. Moreover, RFID will allow real-time inventory control and will support the store managers in the process of placing appropriate replenishment orders. The time required for the quarterly inventory and packing returnable items will fall to only a few minutes, meaning that store personnel have more time to devote to exclusive customer care.

Table 9.1 Current store operations records

Task description	Average record
Inventory receiving	100 items/hour 25% of available staff involved All items managed within the day
Initial store inventory	2200 items
Inventory taking	4 times a year 170 items/person*hour
Price re-labeling for sale period	120 items/person*hour
Store opening time	9.5 hours/day
Full-time employees	2 persons
Part-time employees	3 persons
Salespersons utilization	70%
Shop window dressing time	2 times a month 2 hours/shop window
Preparation of items to be returned	150 items/person*hour 850 items (at the end of each season)
Items per case	45
Quantity of items shipped for each new theme	400
Shipment lead-time	1.5 days
Sales share index	35% half-a-day, Mon-Fri morning 65% half-a-day, Mon-Fri afternoon 50% half-a-day, weekend morning 50% half-a-day, weekend afternoon
Preparation time at the DC	0.5 day

9.5 Modeling of store operations at Miroglio

This section provides a description of the main stages of developing the SD model, namely the causal loop diagram and the numerical SD model built using the Vensim® DSS software tool.

9.5.1 Causal loop diagram

Before examining the main feedback loops, one important assumption of our SD model needs to be discussed: the relationship between the time available to assist customers and sales. A survey carried out in the 850 Miroglio stores in Italy revealed that store managers perceive the time spent by salespersons in assisting customers as a key factor in increased sales, probably because staff-assisted sales are a major part of the company's strategy. This perception is reasonably accurate, since a statistical analysis of historical data collected by the company revealed a slight positive correlation between the time dedicated to helping customers and sales (Fig. 9.1). This finding suggested that the relationship was worth exploring in the study.

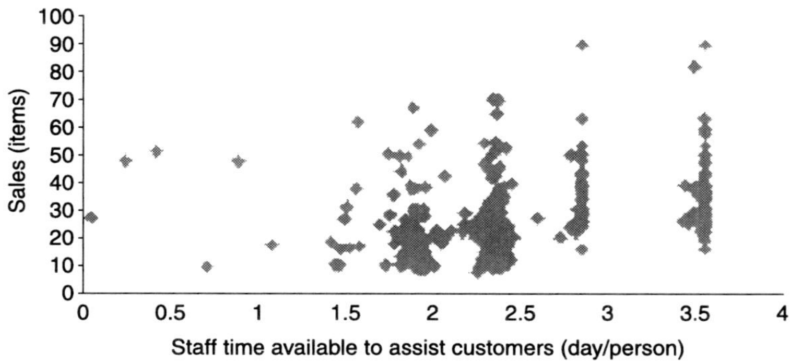

9.1 Correlation between staff time available to assist customers and sales.

Figure 9.2 illustrates the feedback loops that have the greatest impact on the final SD case model. The loops are extracted from the total diagram:

- *Loop No. 1* is a representation of the balancing feedback effect of sales on store inventory. The larger the inventory, the more time staff spend handling items (labeling, counting, etc.), resulting in less time being available to help customers with an associated negative impact on sales. If the sales fall, fewer orders will be placed, leading to fewer items coming into the store and slower inventory growth.
- *Loop No. 2* shows the balancing impact of order rate on store inventory. In fact, the order rate is determined by both the sales forecast, through the reorder point and the current level of inventory. If the level of inventory increases, fewer items will be ordered per unit of time, assuming that the reorder point remains equal. As a consequence, fewer items will enter the store, resulting in a store inventory level decrease.

The feedback structure presented in Fig. 9.2 also includes two other reinforcing loops; Loop No. 3a and Loop No. 3b are somewhat similar. As the store inventory increases, less time will be available to assist customers, so that sales will decrease, with a subsequent reduced rate of outgoing items and, therefore, inventory growth. Finally, it should also be noted that the 'assisted sales index' has been defined to quantitatively represent the impact on sales of the time spent by staff on customer care. This will be discussed in depth in a later section.

9.5.2 SD Model

The model is an illustration of the average behavior of the real system: the large variety of garments, sizes and colors is reduced to one sample item and the variety of points of sale is limited to one sample store. The model makes use of the average values reported in Table 9.1. Figure 9.3 illustrates the most relevant parts

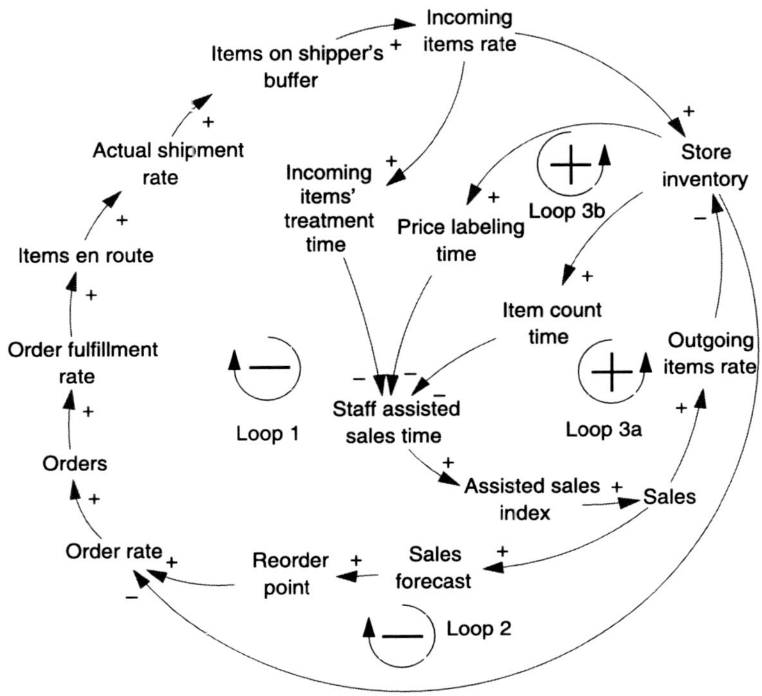

9.2 Main feedback loops interacting in the store operations system.

of the model. The full list of equations is available in the Appendix at the end of the chapter (Section 9.11).

The SD model developed here is a representation of a variable reorder quantity and a stock control policy with a variable review period in times of uncertain demand. It can be broadly divided into two stock and flow sections, namely orders and inventory, and a system of interconnecting variables mainly related to staff usage and sales demand. The top portion of the diagram depicts the order adjustment process: the accumulation of orders is increased by the incoming order rate and decreased by the order fulfillment rate. The rate of orders received at the DC is determined according to

$$\textit{order rate} = \textit{MAX}(0, (\textit{reorder point} - \textit{store inventory} - \textit{items en route})/\textit{day})[\text{Items/Day}] \quad [9.1]$$

It should be noted that the order rate refers to replenishment orders of sold items. This kind of order is different from other types, namely the 2-week central orders and the new-season orders. Two-week central orders are related to the new themes that Miroglio ships to stores every two weeks during a season, while new-season orders are related to the new themes delivered to stores before the season launch.

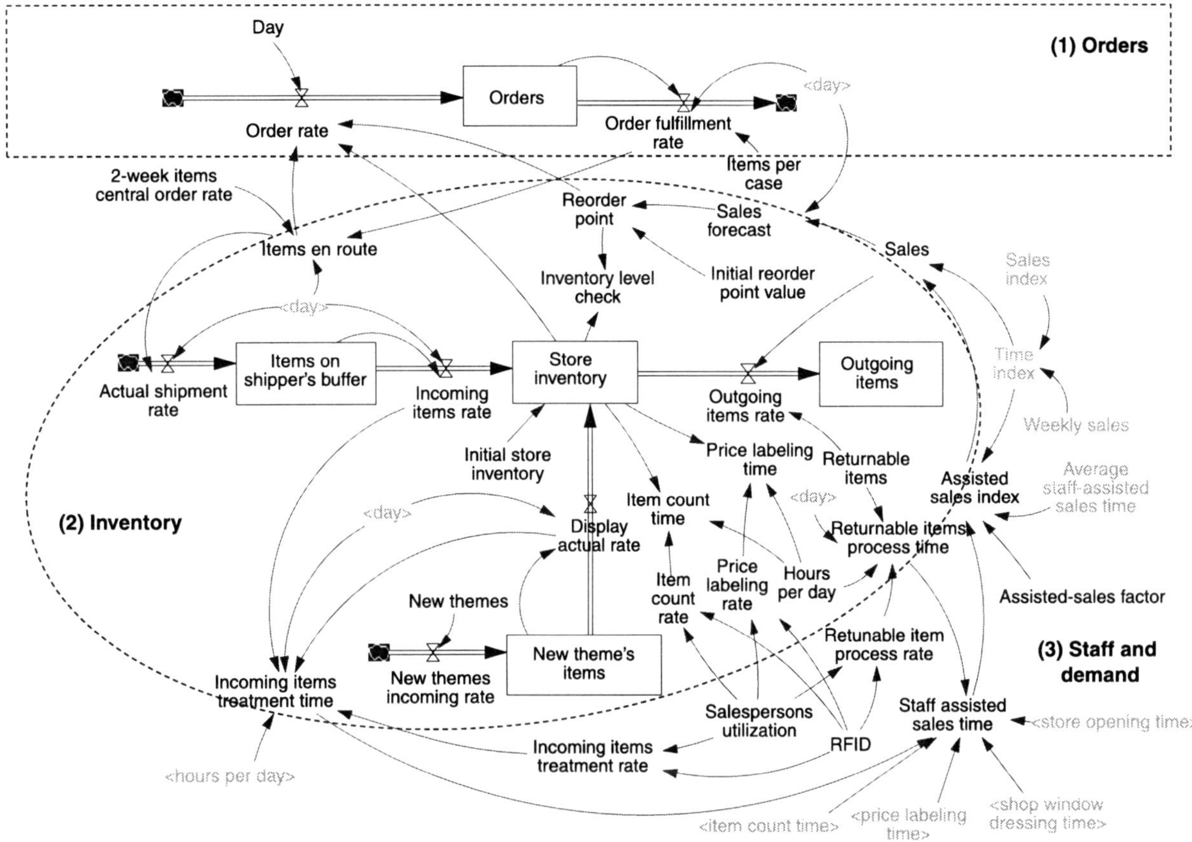

9.3 Portion of the SD model of Miroglio's Fast Fashion store operations.

As a consequence, 'items en route' during a season includes both the '2-week items central order rate' and the 'order fulfillment rate', multiplied by the time. The reorder point is initially determined at the beginning of the season and then adjusted throughout the season by both store managers and the central marketing department based on the sales forecast. This has been modeled with a 14-day simple exponential smoothing (Silver *et al.*, 1998); the selected smoothing period is consistent with the period of time that a theme is displayed and allows stochastic noise to be eliminated. The relevant equation is

$$sales\ forecast = SMOOTH(sales*day,\ 14) \quad \text{[Items]} \qquad [9.2]$$

The portion of the stock and flow diagram that models the store inventory makes use of a dummy incoming buffer (items on shipper's buffer) in order to eliminate shipments on Saturday afternoons and Sundays, such that Friday shipments are simulated to arrive early the following week. The 'incoming items rate' variable refers to those items that enter the store and are promptly displayed for sale. If items that are part of a new theme arrive before the beginning of a season, these are partly stocked in the backroom (new theme's items) for reasons of limited space, and are progressively displayed on the shelf (display actual rate). The store inventory level is initially determined at the beginning of a season and then increased by both the incoming items rate and the display actual rate. The inventory is reduced by both items sold over the course of the season (the 'sales' variable) and items returned at the end of the promotion period (returnable items). It is also worth noting that the variable 'inventory level check' returns a binary value, indicating the time when the level of store inventory is lower than the reorder point and, therefore, a replenishment order needs to be placed. Although this type of variable is not strictly essential for the model, it is introduced here for the sake of model testing (Section 9.3).

The lower right-hand portion of Fig. 9.3 relates to measuring the effects of the various feedbacks on the time that store staff can spend on assisting customers and hence increasing sales. This time is equal to the total available productive working hours minus the time required to receive inventory and prepare items for shelf display, dress shop windows and prepare returnable items. The complete definition is given by

$$\begin{aligned} staff\text{-}assisted\ sales\ time = {} & store\ opening\ time - (shop\ window\ dressing \\ & time + treatment\ time\ for\ incoming \\ & items + returnable\ items\ process\ time + item \\ & count\ time + price\ labeling\ time) \quad \text{[Day]} \qquad [9.3] \end{aligned}$$

Store opening time is determined by adding the product of full-time employees and full-time working hours to the product of part-time employees and part-time working hours. The resulting quantity is then divided by the number of hours per day.

The 'sales' variable is an important element that describes demand and is used to quantify the impact of the increased staff-assisted sales time granted by RFID technology. Sales resulting from the summation of two factors are given as

$$sales = time\ index + assisted\ sales\ index \quad \text{[Items/Day]} \qquad [9.4]$$

The 'time index' variable allows weekly estimated sales based on previous sales records to be included, and is calculated as

$$time\ index = weekly\ sales * sales\ index \quad \text{[Items/Day]} \qquad [9.5]$$

'Weekly sales' is a fourth-order polynomial curve line model that reasonably fits previous series of sales data (Fig. 9.4), and the 'sales index' (Table 9.1) allows sales to be distributed across every week and day of any one season. It is worth noting that the weekly sales variable represents typical seasonality as a trend, rather than as total values.

The assisted sales index shows the influence of assisted sales time on sales performance, according to

$$\text{assisted sales index} = \text{time index} \times \frac{(\text{staff-assisted sales time} - \text{average staff-assisted sales time})}{\text{average staff-assisted sales time}}$$
$$\text{assisted sales factor} \times \text{[Items/Day]} \qquad [9.6]$$

The time index variable is multiplied by the percentage variation of staff-assisted sales time in comparison to the average daily-assisted sales time at each time step of the simulation, and by the 'assisted sales factor'. The assisted sales factor is a constant parameter that can be set from 0 to 1, to represent the influence of staff assistance on purchasing effectiveness and sales growth. Miroglio has assumed that its current value is 0.5, based on statistics that half of all transactions are completely assisted. An assisted sales factor of zero would cause the system to fall into the polynomial demand model.

A binary 'RFID' input variable was also included: a value of 0 means that the model replicates the current store conditions, while a value of 1 denotes a reduction

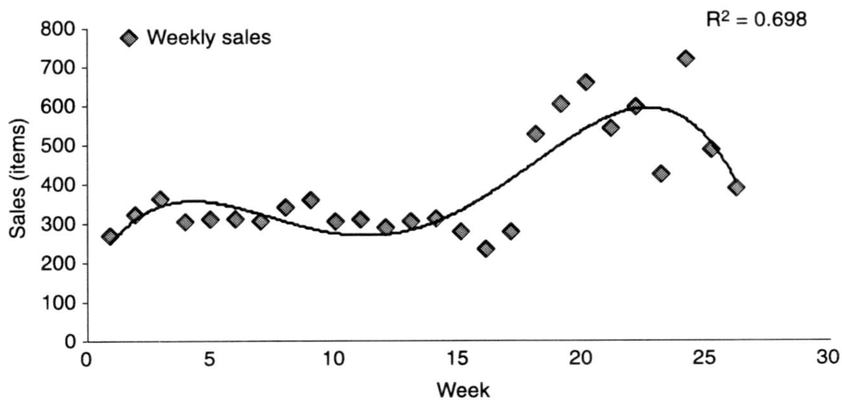

9.4 Scatter plot and fourth-order polynomial fitted curve line of weekly sales.

in the time required to perform backroom operations, and consequently an increased amount of time available for customer care. The model considers the potential of RFID technology to automate and reduce the time required to perform the operations listed in Table 9.1. The simulations were run with Euler integration, a 0.5 time step, a time horizon consistent with a season of 182 working days, steady and unvaried exogenous conditions, and replenishment orders placed up to day 165, just before the beginning of the sale period. Other assumptions are included in Table 9.1.

9.6 Model testing

Before any simulations involving future state scenarios were tested, the first test was to assess whether the simulated results fit the curve lines of previous data sets pertaining to the current scenario in which RFID technology was not used. These tests were taken as an *ex-ante* evaluation of the validity and predictability of the model, and were run by means of curve fitting of the simulated behavior, robustness tests and sensitivity analyses (Sterman, 2000).

The fitting of the model output to real data was checked by focusing on the main variables: 'sales', 'store inventory', and 'actual shipment rate' (replenishment shipments). Figure 9.5 compares the trends of actual versus simulated sales when

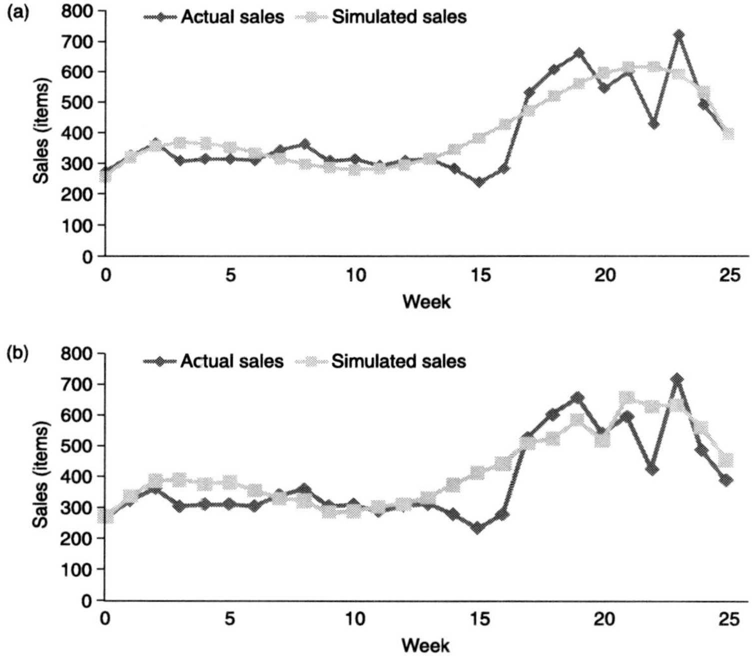

9.5 (a) Actual and simulated sales when the assisted sales factor equals 0; and (b) actual and simulated sales when the assisted sales factor equals 1.

the assisted sales factor variable is set at 0 (Fig. 9.5(a)) and 1 (Fig. 9.5(b)). It is clear that the curve of the simulated outcomes bears a close resemblance to those of the historical outcomes. Only in the last ten weeks of the season do the actual values oscillate more than the simulated values, probably as a result of some particularly appealing characteristics of single themes that are put on the market fortnightly, an aspect that was not considered in our model. Similar results were obtained for the other factors investigated, thus allowing us to conclude that the SD model is able to effectively reproduce the actual behavior of the system at issue.

Model robustness and sensitivity analyses were supported by the specific validation tool developed by Vensim® DSS. Multivariate sensitivity analyses were performed by defining input variables as uniformly distributed and launching 200 simulation runs for each analysis. As a first step, assisted sales factor and sales were assumed to be random input variables. When the level of store inventory is lower than the reorder point, such as when the binary variable inventory level check is activated, it is assumed that there is an increased probability of stock-out, negatively influencing sales. Thus, a sales decrease of 30 to 90% has been assumed in those time steps when inventory level check equals 1. Figure 9.6(a) shows the probability distribution of the store inventory variable, and Fig. 9.6(b) the

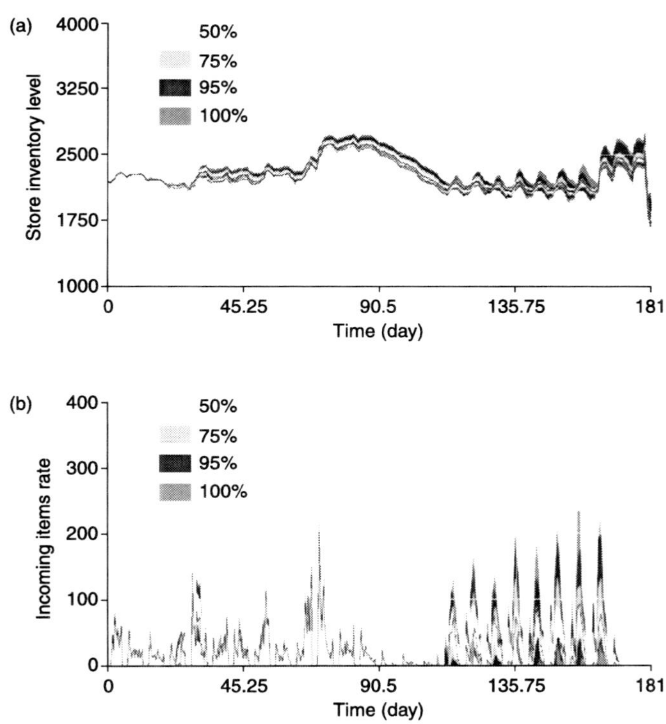

9.6 (a) Probability distribution of store inventory level; and (b) probability distribution of the incoming items rate.

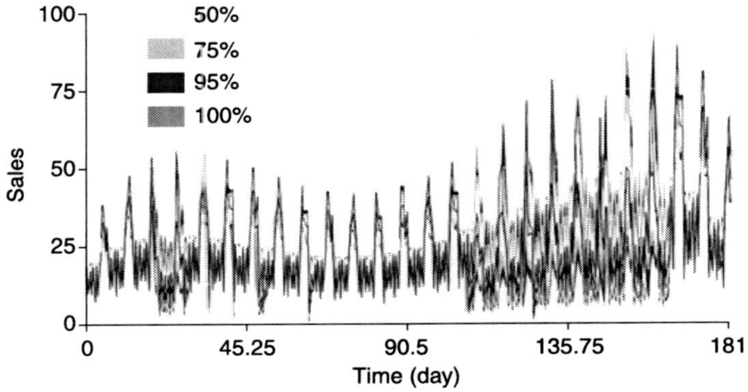

9.7 Probability distribution of sales.

probability distribution of the incoming items rate variable as the percentage of reduced sales varies.

Both variables show similar trends associated with different confidence levels (50, 75, 95 and 100%). The probability distribution of sales provides further evidence of model robustness (Fig. 9.7). First, it should be noted that sales values are never negative. Second, the apparent oscillation observed at the end of the season is not an indication of poor model robustness, but is instead due to the fluctuating market demand during the 8-week-long promotional period.

Sensitivity analyses were also performed on the sales growth data. In this case, when the inventory level check variable is nil, sales are increased by up to 70%. The use of a sales increase scenario and a sales decrease scenario means that the model can be effectively tested for conditions of highly variable demand. Inventory level effects on sales were then removed and the variables 'items per case' and 'salesperson utilization' were defined as random values. Finally, previously defined variations in sales were combined with a scenario in which sales staff are scarce, in order to understand the store behavior predicted by the model under extreme conditions of both highly variable demand and low availability of sales staff. The most relevant stock and flow variables were not found to be sensitive to changes in input quantities.

9.7 Assessing the impact of RFID technology

Assuming an assisted sales factor of 0.5, the simulations indicate that the introduction of RFID technology in Miroglio Fashion stores could increase the time available for staff to assist customers by up to 2.5%, as a result of the reduced time required for backroom operations (Table 9.2). Moreover, if the assisted sales factor increases, the inventory turnover also increases as an associated effect of the increased sales rate. This in turn contributes to a time-based competitive advantage.

Table 9.2 Comparison of time spent on store operations between current status and with RFID technology

	Current [hours]	RFID [hrs] factor 0.5	Variation [%]
Store personnel total working hours	8901.13	8901.13	0.00
Incoming items treatment time	162.85	25.26	–84.49
Inventory taking time	38.63	16.41	–57.52
Sale price relabeling time	38.14	7.69	–79.84
Shop window dressing time	38.38	21.66	–43.56
Returnable items process time	11.67	2.79	–76.09
Staff-assisted sales time	8612.10	8827.95	2.51

The combination of the assisted sales factor with RFID technology directly results in increased sales volumes (Fig. 9.8). The model estimates that seasonal cumulative sales would rise by up to 2.5%, assuming that the assisted sales factor remains constant at 0.5, as a result of the interconnecting dynamic feedback effects of increased assisted sales time, better inventory control, reduced stock-outs and faster inventory turnover. As the assisted sales factor goes up, cumulative sales increase almost linearly. Moreover, the derivatives of curves show that as the assisted sales factor increases, the benefits of RFID technology on sales become more significant. However, when the assisted sales factor is zero, for example when visual merchandising and a self-purchasing strategy are adopted, implementing RFID does not contribute directly to sales growth, but only allows cost savings in labor terms due to shorter backroom operations.

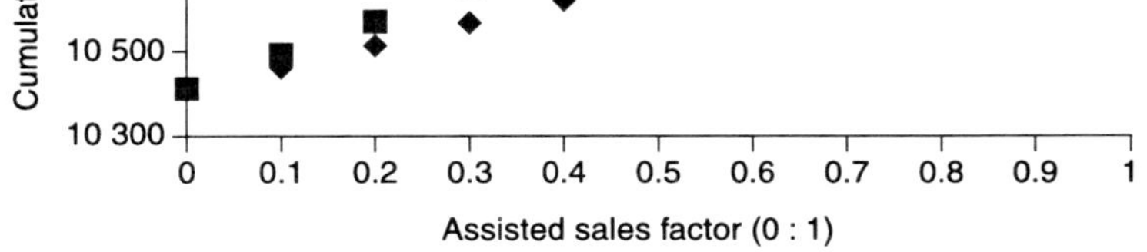

9.8 Growth of sales as a function of the assisted sales factor, in both current and RFID case-scenarios.

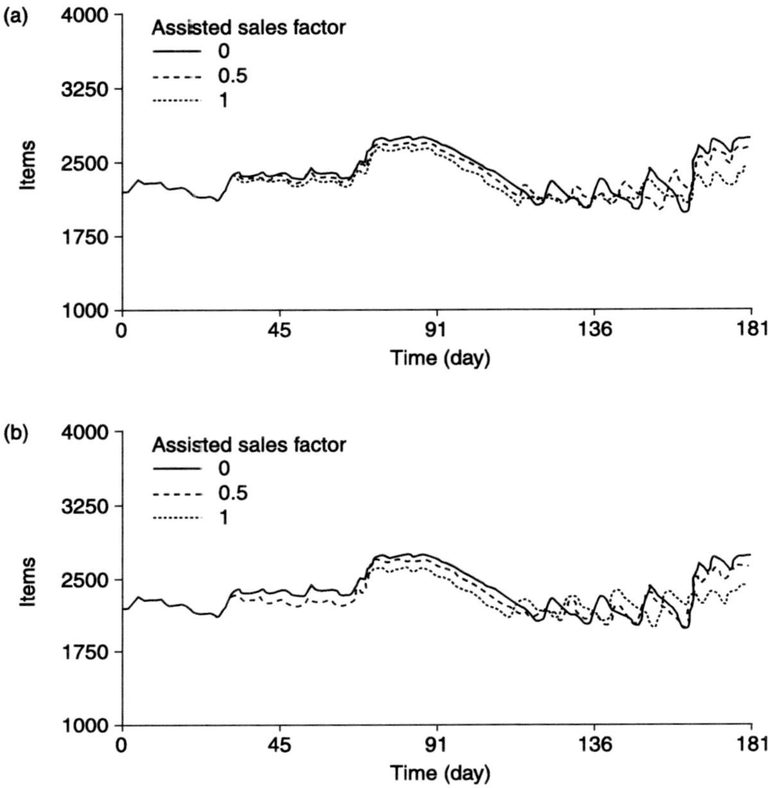

9.9 (a) Inventory level as a function of the assisted sales factor, in the current case-scenario; and (b) inventory level as a function of the assisted sales factor, in the RFID case-scenario.

The quicker inventory turnover permitted by an increased assisted sales factor also allows a lower level of store inventory to be achieved, improving business efficiency. Figure 9.9 shows the inventory level with and without the use of RFID technology against the different values assigned to the assisted sales factor. Without RFID (Fig. 9.9(a)), an increase in the assisted sales factor generally leads to a decrease in inventory level. This benefit is even more significant when RFID technology is employed: the difference between the curve line with an assisted sales factor equal to 1 and the others is greater than the equivalents without RFID (Fig. 9.9(b)).

These results are based on the assumption that the extra time made available by RFID technology is spent serving a greater number of consumers, who can be expected to visit the store when the economy is healthy. Moreover, the longer time available for assisting customers may effectively be used only when the sales strategy is strongly based on staff-assisted sales, as in the case of Miroglio Fashion. However, in periods of low demand, for example, during a stagnating economy, or when sales are not assisted, the time saved in background operations may only lead to cost savings through a reduction in store personnel.

9.8 Assessing return on investment from RFID technology

To complete the process of quantifying the economic benefits of item-level RFID, SD simulation outcomes served as inputs for CFA to evaluate the return on RFID investment achieved by a retailer focused on a salespersons-assisted sales strategy, such as Miroglio. Different market scenarios were considered under different economic conditions. In a stagnating economy, a steady trend of undiminished current sales is assumed over the cash flow period, so that only personnel cost savings are created. In a booming economy, which is modeled using a greater 'sales index' variable, the availability of RFID technology would replace the need for additional personnel to meet the increased demand. Also, with the increased sales index associated with a booming economy, the average inventory level reduces due to a greater sales rate and a shorter inventory turnover, meaning that no extra storage space is required over that currently available (Fig. 9.9(b)).

For each of the two scenarios, 3-year and 5-year net present values (NPVs) and internal rates of return (IRRs) were compared. The discounted pay-back period (DPP) was also computed.

We evaluated the discount rate by using the Weighted Average Cost of Capital (WACC) method, based on Miroglio's current economic performance. Two situations were analyzed: first, the company's investment in RFID technology was assumed to be completely funded by equity; while the second situation assumed a 65% debt leverage, according to Miroglio's standard investment policy. The following additional assumptions were made for the purposes of CFA: a €21 500 investment to equip each individual store with RFID technology; a gross profit per item equal to 75% of sale price, lowered to 37.5% during the promotional period; and a cautious cost estimate of 14€cents per disposable tag. Cash streams did not take into account maintenance and operations costs of RFID middleware, and the effect of taxes was not considered.

With stagnating demand, RFID only allows a reduction in store personnel. The cash flow comprised the initial RFID investment, the cost of disposable tags and the reduced cost of human resources. Personnel cost savings were evaluated by multiplying the cost per hour of individuals by the total amount of hours saved over the course of a month as a result of the reduced time spent on backroom operations. This scenario leads to a DPP of slightly over three years (Table 9.3). Calculations were performed under Miroglio's current value for the assisted sales factor (0.5), since an increased assisted sales factor would not lead to augmented sales due to market stagnating conditions. In addition, the 3-year IRR is lower

Table 9.3 Return on investment under the stagnating economy scenario

Assisted sales factor=0.5						
Leverage	Annual discount rate	Three-year NPV	Five-year NPV	Three-year IRR	Five-year IRR	DPP (months)
100% equity	3.67%	– € 54,335	€ 5,921,685	2.63%	22.25%	37
35% equity-65% debt	3.83%	– € 78,146	€ 5,860,293			37

than the cost of capital. Thus, if the RFID technology became obsolete within three years, the investment should not be pursued.

In a booming economy, the decreased time spent on backroom operations can be used to raise sales. However, in order to ensure that Miroglio stores would actually be able to transform this time into increased sales and would not need to reduce staffing levels, we estimated an extra 5-year promotional investment of €10 500 000 to attract the additional demand required and to achieve the predicted increase in market share. The figure is based on budgeted estimates from the company's marketing department, which indicated that this was a level of effort similar to a previous 5-year plan to increase the market share, so that sales could be expected to increase by 2 to 5% (Fig. 9.8).

The resulting enlargement of the company's market share has been quantified by capturing the raised sales as an increase in the assisted sales factor. Cash flow analyses were performed by setting the assisted sales factor at 0.7, 0.8 and 1, respectively. The first two values represent a medium market share growth, while the third is a strong expansion. Besides the marketing investment, the cash stream also includes the initial investment in RFID technology, increased sales revenues and the costs of tags. Table 9.4 shows the results obtained. The best investment return is achieved when the assisted sales factor equals 1, in line with a large increase in market share, with a DPP lasting for less than two years. However, with a medium growth in market share, the investment is not profitable within a 3-year time frame.

9.9 Conclusion and future trends

This chapter has proposed the use of SD as a structural modeling and simulation method to assist in the process of assessing the economic impact of item-level RFID technology on retail operations, with specific reference to one fast fashion retailer. SD proves to be a valuable empirical approach, because it is able to take into account the complex interrelated feedback effects of the multiple variables involved in overall performance of the system.

Table 9.4 Return on investment under the boosting economy scenario

Leverage	Annual discount rate	Three-year NPV	Five-year NPV	Three-year IRR	Five-year IRR	DPP (months)
Assisted sales factor=0.7						
100% equity	3.67%	– € 6,307,516	€ 860,093	–28.44%	1.06%	59
35% equity-65% debt	3.83%	– € 6,325,667	€ 841,943			59
Assisted sales factor=0.8						
100% equity	3.67%	– € 1,576,702	€ 8,922,477	–3.42%	20.50%	42
35% equity-65% debt	3.83%	– € 1,576,702	€ 8,893,347			42
Assisted sales factor=1						
100% equity	3.67%	€ 8,082,450	€ 24,157,297	34.63%	52.88%	22
35% equity-65% debt	3.83%	€ 8,555,482	€ 24,005,465			22

Using this method, RFID technology has been shown not only to reduce operational costs, but to offer benefits in the revenue stream by allowing increased sales, particularly when the retailer makes use of store personnel to assist customers. RFID technology allows various backroom operations to be automated, with a resulting reduction in the time required to carry out these procedures. It consequently also frees up extra time for the store personnel to assist customers, leading to potential increases in sales and revenue. The case study suggests that retailers focused on salesperson-assisted sales strategies could invest in item-level RFID technology in a booming economy as a promising means of tackling market growth, using the time saved from the automation of backroom operations to take care of a greater number of customers, thus achieving a higher return on the RFID investment. In this context, RFID technology can be used as a driver of competitive advantage and a lever for market share growth.

The integration between SD and CFA adopted in this case-based model can easily be applied to similar retailers, as a tool to provide strategic insights for the implementation of item-level RFID technology. It can also be adopted to evaluate both the operational and economic benefits provided by innovation, wherever industry behavior is determined by a series of complex interconnected feedback effects among relevant variables.

However, it should be noted that the approach suggested in this chapter does have some limitations. First, the assumptions made lead to certain restrictions in applicability: storage costs, any potential RFID technology breakdown-related expenditures, and staff training costs are all ignored. Second, the method disregards potential additional impacts on store operations and sales strategy, such as perpetual inventory and in-store location, theft prevention, check-out automation, interactions with salespeople's productivity, and various innovative marketing technologies, such as dynamic pricing. Moreover, it only takes into account factors relating to product identification, while ignoring possible implications for customer identification and relationship management. Finally, the current study is restricted to just assessing the benefits of RFID on store operations, and does not include any impact on the DC and manufacturing echelons of the SC, nor any additional soft variables involved in restructuring the process. Future research to be carried out by this group should include all of these elements.

It should also be observed that the economic benefits largely depend on the existing technological capacity of the individual retailer, because the lower the current levels of automation, the higher the economic return from RFID implementation. Moreover, the model does not address inter-organizational SC issues. Its main value is therefore for the retailers, who obtain the maximum advantage from in-store implementation of RFID, rather than other SC players (Bottani and Rizzi, 2008). Finally, some limitations are inherent in the predictive nature of the proposed SD decision support model, which would need an *ex-post* validation based on the experience of implementing RFID before it can be assumed to be wholly valid and reliable.

9.10 Acknowledgements

This study is part of Miroglio's SC reengineering project with the involvement of the Engineering Systems and Logistics Research Group (RESLOG) of the Politecnico di Torino. The authors wish to acknowledge Miroglio Fashion S.r.l., Sales and Marketing and Logistics Units, and in particular Piero Abellonio, Logistics Director, and Maurizio Rijllo for active collaboration, support and permission for publication.

9.11 Appendix: full list of the equations of the system dynamics model

"2-week items central order":=GET XLS DATA('file.xls', 'sheet', 'cell', 'cell'); Units: Items/Day

actual shipment rate=(items in route/day); Units: Items/Day

arrival of new season's themes=500 * PULSE TRAIN(163, 0.5, 5, 181); Units: Items/Day

day=1; Units: Day

display actual rate=IF THEN ELSE(PULSE TRAIN(5.2, 1.8, 7, 181), 0, new season's items/day); Units: Items/Day
"full-time employees" = IF THEN ELSE(PULSE TRAIN(4.5, 2.5, 7, 181), 2, 2 * RANDOM UNIFORM(0.65, 0.7, 0.01)); Units: employee
"full-time working hours" = 1*9.5; Units: Hour/employee
hours per day=9.5; Units: Hour/Day
incoming items rate = IF THEN ELSE(PULSE TRAIN(5.2, 1.8, 7, 181), 0, (items on shipper's buffer/day)); Units: Items/Day
incoming items treatment rate=IF THEN ELSE(RFId = 0, (salespersons utilization*RANDOM UNIFORM(95, 105, 1)), salespersons utilization* RANDOM UNIFORM(650, 700, 10)); Units: Items/Hour
incoming items' treatment time = ((incoming items rate + display actual rate) * day / incoming items treatment rate) / hours per day; Units: Day
initial inventory coverage = 2150; Units: Items
intial store inventory = 2200; Units: Items
inventory adjustment = SMOOTH(sales *day ,14); Units: Items
item count rate=IF THEN ELSE(RFId = 0, (salespersons utilization*RANDOM UNIFORM(300, 350, 1))* (PULSE(63, 1) + PULSE(126, 1)), (salespersons utilization *RANDOM UNIFORM (750, 870, 10)) * (PULSE(63, 1) + PULSE(126, 1))); Units: Items / Hour
item count time = IF THEN ELSE(item count rate = 0, 0, (store inventory/item count rate)
/ hours per day); Units: Day
items in route= DELAY FIXED (order fulfillment rate*day + "2-week items central order" * day, (0.5*RANDOM UNIFORM(1.4, 1.7, 0.1 + 0.5* RANDOM UNIFORM(1.9, 2.4, 0.1))), 0); Units: Items
items on shipper's buffer = INTEG (actual shipment rate-incoming items rate,0); Units: Items
items per case = 45; Units: Items
new season's items = INTEG (new themes' incoming rate-display actual rate, 0); Units: Items
new themes' incoming rate=arrival of new season's themes; Units: Items/Day
order fulfillment rate = (MAX (0, INTEGER(orders/items per case) * items per case/day)) * PULSE TRAIN(0, 5, 7, 181); Units: Items/Day
order rate = MAX(0, (reorder point-store inventory-items in route) / day); Units: Items/Day
orders = INTEG (order rate-order fulfillment rate, 0); Units: Items
outgoing items = INTEG (outgoing items rate, 0); Units: Items
outgoing items rate = sales + returnable items; Units: Items/Day
"part-time employees" = IF THEN ELSE(PULSE TRAIN(4.5, 2.5, 7, 181), 3, 3* RANDOM UNIFORM(0.65, 0.7, 0.01)); Units: employee
"part-time working hours" = 0.5*9.5; Units: Hour/employee

price labeling rate = IF THEN ELSE(RFId =0, (salespersons utilization* RANDOM UNIFORM(140, 160, 1)) * PULSE(145, 1), (salespersons utilization*RANDOM UNIFORM(750, 800, 10)) * PULSE(145, 1)); Units: Items/Hour
price labeling time = IF THEN ELSE(price labeling rate = 0, 0, (store inventory/ price labeling rate)/hours per day); Units: Day
reorder point = ACTVE INITIAL ((initial inventory coverage + inventory adjustment)*PULSE(0, 165), initial inventory coverage); Units: Items
returnable items time = IF THEN ELSE(returnable item rate = 0, 0, (returnable items*day/returnable item rate) /hours per day); Units: Day
returnable item rate = IF THEN ELSE(RFId=0, (salespersons utilization*RANDOM UNIFORM(180, 200, 1)) * PULSE(179, 1), (salespersons utilization*RANDOM UNIFORM(750, 800, 10)) * PULSE(179, 1)); Units: Items/Hour
returnable items = RANDOM UNIFORM(700, 800, 10) *PULSE(179, 1); Units: Items/Day
RFId = 1; Units: Dmnl [0,1,1]
sales: = GET XLS DATA('file_motivi.xls', 'vendite', 'A', 'C1'); Units: Items/Day
salespersons utilization = 0.7; Units: Dmnl
shop window dressing rate = IF THEN ELSE(RFId = 0, (RANDOM UNIFORM(2, 2.5, 0.1)/salespersons utilization)* PULSE TRAIN(14, 0.5, 14, 181), RANDOM UNIFORM(1.5, 2, 0.5) * PULSE TRAIN(14, 0.5, 14, 181)); Units: Hour
show window time = IF THEN ELSE(shop window dressing rate = 0, 0, shop window dressing rate/hours per day); Units: Day
"staff-assisted sales time" = MAX(0, store opening time – (show window time + incoming items' treatment time + returnable items time + item count time + price labeling time)); Units: Day
stock out = IF THEN ELSE(reorder point-store inventory >= 0, 1, 0); Units: Dmnl
store inventory = A FUNCTION OF(display actual rate, incoming items rate, initial store inventory, outgoing items rate); Units: Items
store inventory = INTEG (MAX(0, incoming items rate+display actual rate-outgoing items rate), initial store inventory); Units: Items
store opening time = IF THEN ELSE(PULSE TRAIN(0, 0.5, 1.5, 181), 0.8* ((("full-time employees"*"full-time working hours") + ("part-time employees"*"part-time working hours"))/hours per day), ((("full-time employees"*"full-time working hours") + ("part-time employees"*"part-time working hours"))/hours per day)); Units: Day

9.12 References

Akkermans, H. and Dellaert, N. (2005), The rediscovery of industrial dynamics: The contribution of system dynamics to supply chain management in a dynamic and fragmented world, *System Dynamics Review*, 21(3), 173–86.

Anderson E. G. Jr., Morrice, D. J. and Lundeen, G. (2005), The 'physics' of capacity and backlog management in service and custom manufacturing supply chains, *System Dynamics Review*, 21(3), 217–47.

Ashayeri, J. and Lemmes, L. (2006), Economic value added of supply chain demand planning: A system dynamics simulation, *Robotics & Computer-Integrated Manufacturing*, 22(5/6), 550–6.

Bauer, R., Scarborough, D. A. and Stafford, J. (2008), Avery-Dennison Leadership Forum, *RFID in Fashion 2008 Conference*, New York: The Fashion Institute of Technology, 13–14 August 2008. Available from: http://www.rfidjournalevents.com/pres_fashion08.php (accessed 6 March 2009).

Becker, J., Vilkov, L., Weiß, B. and Winkelmann, A. (2010), A model-based approach for calculating the process driven business value of RFID investments, *International Journal of Production Economics*, 127(2), 358–17.

Bottani, E., Ferretti, G., Montanari, R. and Rizzi, A. (2009), The impact of RFID technology and EPC system on logistics processes of the fashion industry supply chain, *International Journal of RF Technologies: Research and Applications*, 1(4), 225–52.

Bottani, E. and Rizzi, A. (2008), Economical assessment of the impact of RFID technology and EPC system on the fast-moving consumer goods supply chain, *International Journal of Production Economics*, 112(2), 548–69.

Cagliano, A. C. (2010), *Understanding Supply Chain Complexity: An Approach Integrating Performance Measurement and System Dynamics*, Saarbrücken, Germany: VDM Verlag Dr. Muller.

Chang, S., Klabjan, D. and Vossen, T. (2010), Optimal radio frequency identification deployment in a supply chain network, *International Journal of Production Economics*, 125(1), 71–83.

Christopher, M., Lawson, R. and Peck, H. (2004), Creating agile supply chains in the fashion industry, *International Journal of Retail & Distribution Management* 32(8), 367–76.

Croson, R. and Donohue, K. (2005), Upstream versus downstream information and its impact on the bullwhip effect, *System Dynamics Review*, 21(3), 249–60.

Curtin, J., Kauffman, R. J. and Riggins F. J. (2007), Making the 'MOST' out of RFID technology: A research agenda for the study of the adoption, usage and impact of RFID, *Information Technology & Management*, 8(2), 87–110.

Delen, D., Hardgrave, B. C. and Sharda, R. (2007), RFID for better supply-chain management through enhanced information visibility, *Production and Operations Management*, 16(5), 613–24.

Disney, S. M., Naim, M. M. and Potter, A. (2004), Assessing the impact of e-business on supply chain dynamics, *International Journal of Production Economics*, 89(2), 109–18.

Doerr, K. H., Gates, W. R. and Mutty, J. E. (2006), A hybrid approach to the valuation of RFID/MEMS technology applied to ordnance inventory, *International Journal of Production Economics*, 103(2), 726–41.

Forrester, J. W. (1961), *Industrial Dynamics*, Cambridge, MA: MIT Press.

Gaukler, G. M., Seifert, R. W. and Hausman, W. H. (2007), Item-level RFID in the retail supply chain, *Production and Operations Management*, 16(1), 65–76.

Ghemawat, P. and Nueno, J. L. (2006), *Zara: Fast Fashion*. HBS 703-497, Boston, MA: Harvard Business School Publishing. Harvard Business Online, available from: http://harvardbusinessonline.hbsp.harvard.edu.

Gonçalves, P., Hines, J. and Sterman, J. (2005), The impact of endogenous demand on push–pull production systems, *System Dynamics Review*, 21(3), 187–216.

Hardgrave, B. C., Langford, S., Waller, M. and Miller, R. (2008), Measuring the impact of RFID on out of stocks at Wal-Mart, *MIS Quarterly Executive*, 7(4), 181–92.

Ju, T. L. (2008), A strategic examination of radio frequency identification in supply chain management, *International Journal of Technology Management*, 43(4), 349–62.

Koh, C. E., Kim, H. J. and Kim, E. Y. (2006), The impact of RFID in retail industry: Issues and critical success factors, *Journal of Shopping Center Research*, 13(1), 101–17.

Koschat, M. A. (2008), Store inventory can affect demand: empirical evidence from magazine retailing, *Journal of Retailing*, 84(2), 165–79.

Kumar, S., Kadow, B. B. and Lamkin, M. K. (2011), Challenges with the introduction of radio-frequency identification systems into a manufacturer's supply chain: A pilot study, *Enterprise Information Systems*, 5(2), 235–53.

Lee, H. and Özer, Ö. (2007), Unlocking the value of RFID, *Production and Operations Management*, 16(1), 40–64.

Lee, I. and Lee, B. C. (2010), An investment evaluation of supply chain RFID technologies: A normative modeling approach, *International Journal of Production Economics*, 125(2), 313–23.

Lee, Y. M. (2009), A quantitative view on how RFID can improve inventory management in a supply chain, *International Journal of Logistics: Research & Applications*, 12(1), 23–43.

Lo, W.-S., Hong, T.-P. and Jeng, R. (2008), A framework of E-SCM multi-agent systems in the fashion industry, *International Journal of Production Economics*, 114(2), 594–614.

Loebbecke, C. (2008), A competitive perspective on standard-making: Kaufhof's RFID project in fashion retailing, *Electronic Markets*, 18(1), 30–8.

Loebbecke, C. and Huyskens, C. (2006), Weaving the RFID yarn in the fashion industry: The Kaufhof case, *MIS Quarterly Executive*, 5(4), 169–79.

Lowson, R. H., King, R. and Hunter, N. A. (1999), *Quick Response: Managing the Supply Chain to Meet Customer Demand*, Hoboken, NJ: John Wiley & Sons, Inc.

Lyneis, J. M. (1998), System dynamics for business strategy: A phased approach, *System Dynamics Review*, 15(1), 35–70.

Masson, R., Iosif, L., MacKerron, G. and Fernie, J. (2007), Managing complexity in agile global fashion industry supply chains, *International Journal of Logistics Management*, 18(2), 238–54.

Mithas, S., Tafti, A., Bardhan, I. and Goh, J. M. (2008), Resolving the profitability paradox of information technology: Mechanisms and empirical evidence. Working paper, College Park, MD: R. H. Smith School of Business, University of Maryland.

Mittelstaedt, R. A. and Stassen, R. E. (1990), Shopping behavior and retail merchandising strategies, *Journal of Business Research*, 21(3), 243–58.

Moon, K. L. and Ngai, E. W. T. (2008), The adoption of RFID in fashion retailing: A business value-added framework, *Industrial Management & Data Systems*, 108(5), 596–612.

Ngai, E. W. T., Moon, K. L., Riggins, F .J. and Yi, C. Y. (2008), RFID research: An academic literature review (1995–2005) and future research directions, *International Journal of Production Economics*, 112, 510–20.

Özelkan, E. C. and Galambosi, A. (2008), When does RFID make business sense for managing supply chains? *International Journal of Information Systems and Supply Chain Management*, 1(1), 15–25.

Pardue, J. H., Clark, T. D. Jr. and Winch, G. W. (1999), Modeling short- and long-term dynamics in the commercialization of technical advances in IT producing industries, *System Dynamics Review*, 15(1), 97–105.

Prater, E. and Frazier, G. V. (2005), Future impacts of RFID on e-supply chains in grocery retailing, *Supply Chain Management*, 10(2), 134–42.

Rust, R. T., Moorman, C. and Dickson, P. R. (2002), Getting return on quality: Revenue expansion, cost reduction, or both? *Journal of Marketing*, 66(4), 7–24.

Silver, E. A., Pyke, D. F. and Peterson, R. (1998), *Inventory Management and Production Planning and Scheduling*, Hoboken, NJ: John Wiley & Sons, Inc.

Sterman, J. D. (2000), *Business Dynamics: Systems Thinking and Modeling in a Complex World*, Boston, MA: McGraw-Hill.

Thiesse, F., Al-Kassab, J. and Fleisch, E. (2009), Understanding the value of integrated RFID systems: A case study from apparel retail, *European Journal of Information Systems*, 18(6), 592–614.

Ton, Z., Dessain, V. and Stachowiak-Joulain, M. (2005), RFID at the METRO Group. HBS 606-053, Boston, MA, Harvard Business School Publishing, Harvard Business Online, available from: http://harvardbusinessonline.hbsp.harvard.edu.

Tzeng, S. F., Chen, W. and Pai, F. (2008), Evaluating the business value of RFID: Evidence from five case studies, *International Journal of Production Economics*, 112, 601–13.

Ustundag, A. (2009), The impacts of Radio Frequency Identification technology on supply chain costs, *Transportation Research Part E – Logistics and Transportation Review*, 45(1), 29–38.

Visich, J. K., Li, S., Khumawala, B. M. and Reyes, P. (2009), Empirical evidence of RFID impacts on supply chain performance, *International Journal of Operations and Production Management*, 29(12), 1290–315.

Whitaker, J., Mithas, S. and Krishnan, M. S. (2007), A field study of RFID deployment and return expectations, *Production and Operations Management*, 16(5), 599–612.

White, A. S. (1999), Management of inventory using control theory, *International Journal of Technology Management*, 17(7/8), 847–61.

Winch, G. W. (1997), The dynamics of process technology adoption and the implications for upgrade decisions, *Technology Analysis & Strategic Management*, 9(3), 317–29.

Index

CPSIA information can be obtained at www.ICGtesting.com
Printed in the USA
BVOW03*1643130414

350230BV00011B/227/P

9 780857 098054